In Search of the Multiverse

In Search of the Multiverse

*Parallel Worlds, Hidden Dimensions,
and the Ultimate Quest for
the Frontiers of Reality*

JOHN GRIBBIN

John Wiley & Sons, Inc.

For general information about our other products and services, please contact our Customer Care Department within the United States at (800) 762-2974, outside the United States at (317) 572-3993 or fax (317) 572-4002.

Wiley also publishes its books in a variety of electronic formats. Some content that appears in print may not be available in electronic books. For more information about Wiley products, visit our web site at www.wiley.com.

ISBN 978-0-470-61352-8 (cloth); ISBN 978-0-470-92656-7 (ebk); ISBN 978-0-470-92657-4 (ebk); ISBN 978-0-470-92658-1 (ebk)

Printed in the United States of America

10 9 8 7 6 5 4 3 2 1

Many and strange are the universes that drift like bubbles in the foam upon the River of Time.

Arthur C. Clarke, 'The Wall of Darkness',
in *The Other Side of the Sky*

If we dismiss theories because they seem weird, we risk missing true breakthroughs. Max Tegmark, MIT physicist

Contents

Acknowledgements

The Multiverse idea is one that has intrigued me since childhood, and one in which I began to take what might be called a scientific interest in the mid-1960s; this book has benefited from discussions and correspondence I have had with various people over more than forty years, as well as those contacted specifically for this project. I cannot remember them all, but among those I cannot forget are Jim Baggott, Julian Barbour, Warwick Bilton, Raphael Bousso, John W. Campbell, Bernard Carr, Louise Dalziel, Paul Davies, Richard Dawkins, David Deutsch, George Ellis, John Faulkner, William Fowler, Neil Gershenfeld, Nicolas Gisin, Simon Goodwin, Philippe Grangier, Ted Harrison, Mark Hindmarsh, Fred Hoyle, Lawrence Krauss, Louise Lockwood, Jim Lovelock, William McCrea, Paul Parsons, Joe Polchinski, Martin Rees, Lee Smolin, Leonard Susskind, Max Tegmark, Edward Tryon, Alex Vilenkin and Ronald Wiltshire. Kenneth Ford and Eugene Shikhovtsev gave permission to quote material from Shikhovtsev's draft biographical sketch of Hugh Everett.

At the other end of the scale of scientific expertise, my writing has recently benefited from the critical comments of my daughter-in-law, Eleanor Gribbin, who has passed through the English educational system almost untouched by science, and is able gently to remind me that it is not always true that 'everybody knows' the things I think they ought to know. And, as with all my books, the input provided by Mary Gribbin has ensured that the end product tells a coherent story instead of presenting a disjointed collection of my favourite flights of fancy.

The metaphor of the Multiverse library came to me during a visit to Clays printers; many thanks to them, both for a splendid day out and for planting the seed of this idea.

Preface: The Search

I have always been fascinated by the relationship between life and the Universe. How did we get to be here? Why is the Universe so big? How did it all begin – and how will it all end? My search for answers to these questions led me to learn as much as I could about astronomy and cosmology, quantum physics, evolution, the history of the Earth and the possibility of life on 'other earths' elsewhere in the Universe. One of the most important things I learned along the way is that our Universe is governed by a set of very simple laws which not only allow but demand the growth of the kind of complexity that has led to the emergence of complex things like us. But I also learned to think of 'our Universe' rather than 'the Universe', because there seems to be no reason why different sets of rules could not operate in different regions of spacetime, beyond the boundaries of anything we could ever see, producing universes different from ours, in which, perhaps, complex things like people could not emerge.

This is the idea of the 'Multiverse'. It is often, though not always, associated with the idea of 'anthropic cosmology', which suggests that we see around us a Universe just right for life because there is a multitude of universes with different physical laws, and life forms like us can only exist in universes like ours. The others are sterile, so there is nobody there to notice what their strange laws of physics are.

So where are these other universes? We cannot search for them physically, like James Cook searching for the southern continent by sailing around the globe in the eighteenth century, but mathematicians and physicists are searching for them metaphorically, developing equations and computer models to try to describe the Multiverse. In fact, they have discovered many different kinds of Multiverse. The

Multiverse may be infinite in space, so that regions of space with different physical laws are separated by infinite distances and can never communicate with one another. The Multiverse could be infinite in time, so that different universes with different physical laws are strung out, one after the other, like beads on a wire, and can never communicate with one another. The Multiverse could contain an infinite number of universes separated in different dimensions, like pages in an infinitely thick book, with each page representing a universe that is unable to communicate with the other pages in the book. And there are other possibilities.

This book is about the metaphorical search for the Multiverse, and the variety of possibilities being considered today. It is impossible to say which, if any, of these ideas is 'right', in the sense of providing an accurate description of the real world, but, as will become clear, I do have my own favourite. It will also become clear, I hope, that there may be less difference between these views of the Multiverse than meets the eye; all of them may be important to an understanding of the Multiverse. The most important thing, a dramatic change that has occurred over the past twenty years or so, is that such ideas are now treated seriously within the scientific community, and are no longer seen as the wild-eyed imaginings of theorists who have been reading too much science fiction. There is a growing body of evidence, increasingly difficult to ignore, that there really is more to the world than the Universe we can see directly.

Even if none of the Multiverse ideas now being investigated by theorists proves to be right – even if it is impossible to prove them right or wrong – this is a shift of perspective as profound as the one which displaced the Earth from the notional centre of the Universe. That is a claim that has been made so often that it has become a cliché; this time, it's true. It even turns out to be relevant to the question of whether the Universe appeared by accident or design, suggesting an answer that was not what I was expecting when I started looking at the evidence. There are still more questions than answers associated with the concept of the Multiverse, but it is definitely time to see what those questions are, and how the search for answers is progressing.

Introduction: In an Infinite Universe, Anything Is Possible

Five hundred years ago, the Universe seemed to be a small place. It was widely believed that the Earth, our home, was the most important thing in the Universe, and lay at its centre. The Sun and the five known planets (Mercury, Venus, Mars, Jupiter and Saturn) were thought to be relatively small objects in orbit around the Earth, and the stars were regarded as points of light attached to a spherical shell around the Earth, rotating once a day just beyond the orbits of the planets. Apart from the rhythms of day and night and the seasons, this setup seemed to be unchanging and eternal. The thought of there being other worlds was literally heretical. Right at the end of the sixteenth century, Giordano Bruno was burned at the stake for espousing ideas which ran counter to mainstream Catholic teaching. These included the idea that the stars are other suns, and that there must be other earths, and life elsewhere in the Universe – although this particular belief was not the main reason for his conviction.

Things began to change with the work of Nicolaus Copernicus, whose famous book *Die Revolutionibus Orbium Coelestium* was published in 1543. Even in ancient times there had been philosophers who speculated that the Earth might go around the Sun, but these ideas never gained widespread acceptance before the sixteenth century. It was Copernicus who started the unbroken line of investigation that has led to our present understanding of the Universe. The shock of his ideas was not just the suggestion that the Earth must move in order to orbit the Sun, but the implication that the Earth is just one of the Sun's retinue of planets. The other planets might be just as important in the cosmic scheme of things as our home in space.

The next shock was the suggestion that the Sun might not be the most important object in the sky, but just an ordinary star. In England in 1576, Thomas Digges, after looking at the Milky Way through a telescope and seeing a multitude of stars, stated in a book called *Prognostication Everlasting* that the Universe is infinite, with stars extending in all directions; Bruno picked up these ideas when he was in England in the 1580s. Galileo Galilei and Johannes Kepler built on the work of Copernicus, and in the seventeenth century astronomers began to estimate the distances to the stars by assuming that they are each as bright as our Sun, but only look faint because they are so far away. In 1728, Isaac Newton came up with an estimate that the star Sirius is about a million times farther from us than the Sun is, not hopelessly far from the distance measured by modern techniques. By then, with the true nature of the orbits of the planets understood, the distances to the Sun and the planets had been determined using geometrical techniques, and astronomers knew that the Sun is about 150 million km from Earth (149,597,870 km according to modern measurements). Saturn, the most distant planet from the Sun known to the Ancients, is nearly ten times farther from the Sun than we are. In the span of just two hundred years, what had seemed to be the entire Earth-centred Universe had shrunk in the considerations of astronomers to a small corner of a vast, possibly infinite, Universe.

It took another two hundred years for these ideas to be assimilated, and for the technologies of telescopes, astronomical photography and spectroscopy to be developed to the point where the next big leap could be made. Along the way, the discovery of more planets in the Solar System beyond the orbit of Saturn (Uranus and Neptune) mattered less for the big picture than the development of accurate techniques for measuring the distances to the stars and the revelation from spectroscopy of what stars are made of. From the 1920s onwards, these techniques led first to a better understanding of our place in the geography of the Universe, and then to a better understanding of our place in the history of the Universe.

With his small telescope, Thomas Digges had seen the band of light we call the Milky Way to be made up of innumerable stars. Galileo, who didn't know about Digges' work, made the same discovery independently a few decades later. Digges thought that the array of stars revealed by the telescope extended to infinity in all directions; but as early as 1750 the Durham astronomer Thomas Wright argued in his book *An Original Theory or New Hypothesis of the Universe* that the way the Milky Way forms a band of light across the sky implies that it is a disc-shaped system with a finite size, which he described as being like the grinding wheel of a mill. Crucially, he realized that the Sun is not at the centre of this slab of stars. He also suggested that fuzzy patches of light revealed by telescopes and known as nebulae lie outside the Milky Way.

Wright's theoretical reasoning was way ahead of its time, and could not be tested by observations with the technology of the eighteenth and nineteenth centuries. His work was largely forgotten until after the observations, made in the twentieth century, which showed that the Milky Way matches the broad outline of his speculations but provided much more precise detail about the nature of the Universe we live in.

Starting from observations made in the 1920s, we now know that the Milky Way is indeed a roughly disc-shaped system, containing hundreds of billions of stars each broadly similar to our Sun, held together by gravity and orbiting their common centre. The disc is about 100,000 light years across (roughly 30 kiloparsecs, in the units favoured by astronomers), so that light, travelling at a speed of just under 300,000 km per second, takes 100,000 years to cross the disc. A light year is about 9.5 thousand billion km. The Sun is roughly two thirds of the way out from the centre of the Milky Way, lying in the plane of the disc, which is about a thousand light years (some 300 parsecs) thick in the vicinity of the Sun. But these impressive statistics, far exceeding the pre-Copernican idea of the Universe, pale almost into insignificance compared with the discovery that the entire Milky Way galaxy is just one island in space, as unspectacular and ordinary a member of the class of galaxies as the Sun is an unspectacular and ordinary member of the class of stars.

Like his ideas about the nature of the Milky Way itself, Wright's speculation that the nebulae – or at least, some of them – lie beyond the Milky Way has also proved correct. Although some nebulae are simply glowing clouds of gas and dust within the Milky Way, and are still referred to by that term, the 'external' nebulae are what we now call galaxies. Galaxies come in different shapes and sizes, but the

Milky Way is an almost exactly average-sized member of the class known as disc galaxies. This is the ultimate recognition of our place in the geography of the Universe – we orbit an ordinary star, one among hundreds of billions of stars in an ordinary galaxy, one among hundreds of billions of galaxies. There is nothing special about our place in the Universe.

It is estimated that there are hundreds of billions of galaxies visible in principle to present-day telescopes, although only a few thousand have been studied systematically. They are distributed in groups known as clusters across the visible Universe, and the most distant yet photographed are seen by light which has spent well over ten billion years on its journey from them to our telescopes. This is not quite the same thing as saying that these galaxies are more than ten billion light years away, because the other great discovery that built from observations made in the 1920s is that clusters of galaxies are moving apart from one another. The Universe is expanding, so after ten billion years the galaxies that emitted that light are no longer at the same distance from us that they were when the light set out on its journey.

This universal expansion is the key to understanding our place in cosmic history. It was discovered by accident, but actually predicted by Albert Einstein's general theory of relativity, although he had ignored the prediction. In the late 1920s and early 1930s, the American astronomer Edwin Hubble was interested in measuring the distances to galaxies, and, working with his colleague Milton Humason, he discovered that the distance to a galaxy is proportional to the redshift of features in its spectrum. This redshift is just what its name suggests – a shift in the spectral features towards the red end of the spectrum, compared with the position of those features measured in the laboratory. Hubble didn't care why the redshift occurred, and didn't try to explain it – he was only interested in using it to measure distances. But it was soon appreciated by other astronomers that the effect is caused by the space between galaxies (strictly speaking, between clusters of galaxies) stretching as time passes.

The reason why the cosmological redshift was interpreted in this way so quickly is that just such a space-stretching effect is a natural consequence of the general theory of relativity, which Einstein developed in the second decade of the twentieth century. At that time, most people still thought that the Milky Way was the entire Universe, and the Milky Way is certainly not expanding. So Einstein had introduced an extra term into his equations, denoted by the Greek letter lambda (Λ) and often referred to as the 'cosmological constant'. When the constant was removed, the equations of the general theory naturally predicted a universe* expanding in exactly the way that the observations of galaxies revealed. It is important to appreciate that this expansion is indeed caused by the stretching of space itself. The cosmological redshift is not caused by the galaxies moving through space, although it is possible to produce redshifts in that way (and blueshifts as well) through the Doppler effect. Red light has longer wavelengths than blue light, and the cosmological redshift is caused by light waves being stretched on their journey to us as the space between us and the distant galaxy stretches.

A key feature of this kind of expansion is that it has no centre, unlike the way fragments of shrapnel move outwards from the site of an exploding bomb. By analogy with the Doppler effect, astronomers often refer to the 'recession velocity' of a galaxy, even though they know the cosmological redshift is not caused by the galaxy moving through space. What they really mean is, the *equivalent* velocity that would be required to produce the same redshift by the Doppler effect. In that language, the velocity is proportional to the distance from us – but you would see the same thing, velocity proportional to distance, from any galaxy. We are not at the centre of the Universe, and there is no centre.

A simple analogy makes this clear. Imagine a sphere, like a basketball, dotted with random spots of paint. If the size of the sphere doubles, every spot of paint will seem to move away from its neighbours. From whichever spot you choose to measure, the other spots will seem to be receding. This is another example of the fact that we occupy no special place in the Universe. The Earth, far from being in a privileged position, seems to be located in such an average place that

---

* I use the term 'Universe' to refer to the totality of everything we can in principle ever see. The term 'universe' is used to refer to mathematical descriptions (models) of the possible behaviour of regions of spacetime, and to refer to other worlds which may exist beyond our space and time. The expanding universe of the general theory is a model, but it matches the behaviour of the real Universe.

the Russian cosmologist Alex Vilenkin has coined the term 'terrestrial mediocrity' to describe our situation.

But although the cosmological redshift is not caused by the galaxies fleeing from one another through space as if from the seat of a great explosion, if we imagine reversing the cosmic expansion we see that at face value the discovery of the cosmological redshift implies that long ago everything we can see around us was compressed into a much smaller volume of space. This is confirmed by the same equations which describe the present expansion of the Universe so well. Modern measurements of the expansion of the Universe, combined with those mathematical models, suggest that the entire visible Universe emerged from a hot fireball of energy occupying a volume smaller than an atom 13.7 billion years ago.

The precision of this number is remarkable. As recently as twenty years ago, cosmologists argued about whether this 'age of the Universe' was nearer to 10 billion years or nearer to 20 billion years, and pinning the number down to within a factor of two seemed impressive enough to anyone not involved in the sometimes heated debate. Now, there is very little room for manoeuvre, and the age of the Universe seems definitely to lie between 13.6 and 13.8 billion years. The outburst of the Universe from this tiny origin is known as the Big Bang, a term coined by the British cosmologist Fred Hoyle to poke fun at what he saw as a ludicrous idea, but one which has stuck – even though it was not an explosion and nothing went bang.

The discovery of the finite age of the Universe, combined with evidence that the Universe changes ('evolves') as time passes, puts us in our place in cosmic history. It turns out that in a sense we do live at a special time in history, although that does not conflict with the idea (sometimes called a 'principle') of terrestrial mediocrity. As I shall explain later, it took time for stars and galaxies to evolve and for the chemical elements that we are made of to be processed inside stars. The Sun and the Earth are about 4.5 billion years old, so they were born roughly nine billion years after the Big Bang. This was just about as soon as it would have been possible for planets like the Earth, rich in the chemical constituents of life, to have formed. In that sense, the Earth formed at a special time; but there is no reason to think that it was the only such planet to form.

Which brings us back to Giordano Bruno. Bruno's vision was of a universe filled with an infinite array of stars, each one like the Sun, and with other planets, many of them bearing life. He wrote:

In space there are countless constellations, suns and planets; we see only the suns because they give light; the planets remain invisible, for they are small and dark. There are also numberless earths circling around their suns, no worse and no less than this globe of ours.

It was an early example of a 'many worlds' hypothesis, using the term 'world' in one of its many alternative meanings as a synonym for 'planet'. At the beginning of the seventeenth century, there seemed to be no fundamental reason why those other worlds – all of them – could not, in principle, be seen from Earth, if you had a powerful enough telescope, or even visited, if you had the patience for a very long journey. The fact that stars are grouped together in galaxies like the Milky Way does not affect the thrust of Bruno's argument; but he did not know that the Universe as we know it was born at a finite time in the past, nor that the speed of light is finite. Those facts do alter our perception of what is meant by 'many worlds', and whether they might be observable.

The finite nature of the speed of light was only established in the second half of the seventeenth century, by the Dane Ole Rømer, from observations of eclipses of the moons of Jupiter; the fact that the speed of light is the ultimate speed limit, and nothing can travel faster than light, was only established by Albert Einstein at the beginning of the twentieth century. Since the Universe only came into existence 13.7 billion years ago, there has only been time for light to travel a finite distance through space since the Big Bang. This distance is not 13.7 billion light years, because, as we have seen, space has been expanding while light has been on its journey. For this reason, careful astronomers prefer to use the term 'look back time' rather than 'distance' to a particular galaxy. But this does not affect the argument. From our perspective, we can only look out into the Universe, even if we had perfect telescopes, to distances corresponding to a look back time of 13.7 billion years. Light from farther away in the Universe has not yet had time to reach us. But it might! If there are any intelligent observers around on Earth in a billion years from now, they will be able to see out to distances corresponding to a look back time of 14.7 billion years. The bubble of space which can in principle be known to us, and which can in principle affect us, is growing all the time.

The same is true for bubbles of space centred on any galaxy in the Universe. If the Universe is indeed infinite, there may be an infinite number of these bubbles, some overlapping one another, others completely separate from one another, but all inhabiting the same Universe of space and time. There may indeed be an infinite number of worlds, in the sense Bruno would have understood, but no single observer would ever be able to know them all.

And the Universe really could be infinite, even though we see only a finite volume. When cosmologists talk about the origin of the Universe in a fireball of energy smaller than an atom, they mean the entire *visible* Universe. The original superdense state may itself have been infinitely large, and our visible Universe may represent just one tiny piece of that infinite region that swelled up to a much larger size, as I discuss in Chapter Five.

An infinite number of worlds allows for an infinite number of variations and, indeed, an infinite number of identical copies. In that sense, in an infinite Universe, anything is possible, including an infinite number of other Earths where there are people identical to you and me going about their lives exactly as we do; and an infinite number of other Earths where you are Prime Minister and I am King. And so on. But the chances of any of these similar Earths occupying 'our' bubble are vanishingly small. The nearest 'other you' is likely to live in a bubble so far away that, according to a calculation made by the American cosmologist Max Tegmark, to express it in metres you would need a number with 10^{29} zeroes – not 29 zeroes, but 10 *raised to the power of 29* zeroes. For comparison, the total number of atoms in all the stars and galaxies of the visible Universe is estimated to be merely10 raised to the power of 80 – a 1 followed by 80 zeroes.

If that was all there was to the idea of the multiverse, there would be no point in writing this book. But there is more – much more. Arguably, other bubbles within the same expanding space and time do not even count as other universes; they are simply inaccessible parts of our Universe. The true multiverse idea strikes at the heart of our understanding of science, addressing puzzles such as the reason why the laws of physics are the way they are, and why the Universe is a comfortable home for life. The second of those questions, in particular, stimulated a debate in which the term 'multiverse' was first used in an astronomical context, just over a hundred years ago. But the word has since been used with different meanings, and it is important to clear up what I mean by it before proceeding further.

According to the *Oxford English Dictionary*, the word 'multiverse' was first used by the American psychologist William James (the brother of novelist Henry James) in 1895. But he was interested in mysticism and religious experiences, not the nature of the physical Universe. Similarly, although the word appears in the writings of G. K. Chesterton, John Cowper Powys and Michael Moorcock, none of this has any relevance to its use in a scientific context. From our point of view, the first intriguing scientific use of the word followed from an argument put forward by Alfred Russel Wallace, the man who came up with the idea of evolution by natural selection independently of Charles Darwin, that 'our earth is the only inhabited planet, not only in the Solar System but in the whole stellar universe.' Wallace wrote those words in his book *Man's Place in the Universe*, published late in 1903, which developed ideas that he had previously aired in two newspaper articles. Unlike Darwin, Wallace was of a religious persuasion, and this may have coloured his judgement when discussing 'the supposed *Plurality of Worlds*'.* But as we shall see, there is something very modern about his approach to the investigation of the puzzle of our existence. 'For many years,' he wrote:

I had paid special attention to the problem of the measurement of geological time, and also that of the mild climates and generally uniform conditions that had prevailed throughout all geological epochs, and on considering the number of concurrent causes and the delicate balance of conditions required to maintain such uniformity, I became still more convinced that the evidence was exceedingly strong against the probability or possibility of any other planet being inhabited.

This was the first formal, scientific appreciation of the string of coincidences necessary for our existence; in that sense, Alfred Russel

* His emphasis.

Wallace should be regarded as the father of what is now called 'anthropic cosmology'.

Wallace's book stirred up a flurry of controversy, and among the people who disagreed publicly with his conclusions were H. G. Wells, William Ramsay (co-discoverer of the inert gas argon), and Oliver Lodge, a physicist who made pioneering contributions to the development of radio. It was Lodge who used the term 'multiverse', but referring to a multitude of planets, not a multitude of universes.

In scientific circles, the word was forgotten for more than half a century, then invented yet again by a Scottish amateur astronomer, Andy Nimmo. In December 1960, Nimmo was the Vice-Chairman of the Scottish branch of the British Interplanetary Society, and was preparing a talk for the branch about a relatively new version of quantum theory, which had been developed by the American Hugh Everett. This has become known as the 'many worlds interpretation' of quantum physics, with 'world' now being used (as it will be from now on throughout this book) as a synonym for 'universe'. But Nimmo objected to the idea of many universes on etymological grounds. The literal meaning of the word universe is 'all that there is', so, he reasoned, you can't have more than one of them. For the purposes of his talk, delivered in Edinburgh in February 1961, he invented the word 'multiverse' – by which he meant *one* of the many worlds. In his own words, he intended it to mean 'an apparent Universe, a multiplicity of which go to make up the whole . . . you may live in a Universe full of multiverses, but you may not etymologically live in a Multiverse of "universes".'

Alas for etymology, the term was picked up and used from time to time in exactly the opposite way to the one Nimmo had intended. The modern usage of the word received a big boost in 1997, when David Deutsch published his book *The Fabric of Reality*, in which he said that the word Multiverse 'has been coined to denote physical reality as a whole'. He says that 'I didn't actually invent the word. My recollection is that I simply picked up a term that was already in common use, informally, among Everett proponents.' In this book, the word 'Multiverse' is used in the way Deutsch defines it, which is now the way it is used by all scientists interested in the idea of other worlds.* The Multiverse is everything that there is; a universe is a portion of the Multiverse accessible to a particular set of observers. 'The' Universe is the one we see all around us. And where better to start our search for the Multiverse than with Everett himself?

* I refer any offended etymologists to the comment of Humpty Dumpty in *Through the Looking Glass*: 'When I use a word,' Humpty Dumpty said, in a rather scornful tone, 'it means just what I choose it to mean, neither more nor less.'

I

The Coming of the Quantum Cats

Quantum physics is the set of laws that govern the behaviour of things on small scales – essentially, the size of atoms and smaller. To put that scale in perspective, it would take roughly ten million atoms lined up side by side to stretch across the gap between two of the points in the serrated edge of a postage stamp. At one level, it is not surprising that the laws of physics that operate at such small scales are different from the laws that operate on a human scale, which were discovered by Isaac Newton in the seventeenth century. Newtonian physics describes the behaviour of things like billiard balls rolling across a table and colliding with one another, waves rippling across the surface of a pond, or the launch of a rocket ship on its way to Mars. But at another level, it is utterly astonishing that quantum physics turns out to be dramatically different from Newtonian physics – different in its very nature, not just in minor ways. Because, after all, things like billiard balls, water in a pond, and rocket ships are all made of atoms. How can the whole behave so differently from the sum of its parts?

There is no single satisfactory answer to that question. There are several possible answers, all equally valid, which is an unsatisfactory situation in itself. And none of those answers 'make sense' in terms of our everyday experience of the world. This is the single most important thing to take on board concerning quantum physics. It is totally outside our everyday experience. There is no way that the human mind can understand what quantum entities such as light or electrons 'really are'. All we can do is carry out experiments, and interpret the results of those experiments by making analogies with things we think we understand in the everyday world.

NEITHER WAVE NOR PARTICLE

In some experiments, the behaviour of light seems to be like the behaviour of ripples on a pond; in other experiments, it seems to be like a stream of tiny billiard balls. But this does not mean that light 'is' a wave or 'is' a particle, nor even that it is a mixture of wave and particle. It is something we cannot envisage, which if asked one question will respond like a wave, while if asked another question will respond like a particle. The same is true of electrons and all other quantum entities. Perhaps, limited by our human experience, we are asking the wrong questions. But we are stuck with the questions and answers we've got.

As long ago as 1929, the physicist Arthur Eddington summed the situation up, in his book *The Nature of the Physical World*. 'No familiar concepts can be woven around the electron,' he wrote; 'something unknown is doing we don't know what.' And he points out that 'I have read something like it elsewhere –

> The slithy toves
> Did gyre and gimbal in the wabe'*

In this regard, nothing has changed in the past eighty years. We still don't know what electrons (or other quantum entities) are, nor how they do the things they do.

Indeed, by breaking our mental link with things like waves and particles, it might be more helpful to translate all of quantum physics into the language of 'Jabberwocky'; it would certainly make just as much sense. Which makes it all the more remarkable that without knowing what quantum entities are or how they do the things they do, by knowing that they do do certain things when prodded in certain ways physicists are able to use quantum entities. This is a bit like the way you can learn to drive a car by learning how to manipulate the controls, without having the faintest idea what is going on under the bonnet.

To take just two examples, quantum physics is essential for the

* Eddington is quoting from Lewis Carroll's *Jabberwocky*.

design of computer chips, which are in everything from your mobile phone to supercomputers used in weather forecasting, and quantum physics explains how large molecules like DNA and RNA, the molecules of life, work. Studying quantum physics is not just an esoteric hobby for unworldly boffins: it has direct, practical benefits. In this book, however, I am more concerned (except in part of Chapter Three) with the esoteric and, if you like, philosophical implications of quantum theory. And nothing could be stranger than the story of Schrödinger's cat and her successors. But before we meet the felines, there are some basics of quantum physics to come to terms with.

It's a sign of how inadequate our everyday experiences are as a guide to the quantum world that, having cautioned you that quantum entities are neither waves, nor particles, nor a mixture of wave and particle, the best way to begin to get some insight into what goes on at the sub-atomic level is to consider the ways in which such entities behave *like* waves or particles. This provides at least some insight into one of the most important, but also non-commonsensical, features of the quantum world – uncertainty.

A QUANTUM OF UNCERTAINTY

In quantum physics, uncertainty is a precise thing. For a quantum entity, there are pairs of parameters, known as conjugate variables, for which it is impossible to have a precisely determined value of each member of the pair at the same time. The more accurately you know property A, the less accurately you know property B, and vice versa. This is not the fault of our inadequate measuring equipment. It is a law of nature, discovered by the physicist Werner Heisenberg in 1927, and known as Heisenberg's Uncertainty Principle. The most important of these pairs of conjugate variables are position/momentum, and energy/time.

The position/momentum relationship is the archetypal example described by Heisenberg. Momentum, in this context, is equivalent to velocity, and velocity describes both the speed and direction that something is moving in. Heisenberg found that the uncertainty in the position of an entity such as an electron, multiplied by the uncertainty in its momentum, is always bigger than a certain (tiny!) number, Planck's constant, divided by 2π. In principle, you can get as near to this limit as you like. But the more precisely the position of, say, an electron is pinned down, the more uncertainty there is about where the electron is going. The more accurately its momentum (or velocity) is determined, the less accurately is its position defined. This uncertainty is a property of the electron (or other quantum entity) itself. An electron itself does not 'know' both where it is and where it is going at the same time.

This is where the wave and particle analogies are useful. But remember that they are only analogies. A wave is a spread out thing. It might well be travelling in a definite direction at a definite speed, but it cannot be located at a point. A particle, if it is small enough, can very nearly be located at a point, provided that it is not moving with a well-defined momentum. But if it moves – if it has a certain momentum – it is no longer located at a point. The more a quantum entity is constrained by circumstances to act like a wave, the less certain it is where the entity is located. The more it is constrained to act like a particle, the less certainty there is about where it is going.

The standard way to describe this is in terms of probabilities. If an electron is fired off from an electron gun in the direction of a phosphorescent screen, as in an old-fashioned TV cathode ray tube, the moment the electron leaves the gun the wave representing it begins to spread out through space, because its position is uncertain. In principle, the laws of quantum physics tell us, the electron could end up anywhere in the Universe; but there is a very high probability that it will strike the screen and make a spot of light there. The instant it does so, the uncertainty in its position shrinks dramatically to the size of the spot on the TV screen. This is called the collapse of the wave function. Then, the wave begins to spread out again from the new location. Unless the electron has got tied up in an atom, or trapped in some other way, its position becomes more and more uncertain as time passes. If it has got tied up in an atom, it is still constrained by quantum uncertainty; but that is not directly relevant to the search for the Multiverse.

THE ONLY MYSTERY

All this is hard to get your head round. But the essence of the quantum world can be summed up in terms of one simple experiment, involving sending light, or a beam of electrons, through two holes in a blank obstruction. Richard Feynman, who won a Nobel Prize for his work on quantum theory, said that this experiment 'has in it the heart of quantum mechanics. In reality, it contains the *only* mystery.'* And if you find you still have trouble getting your head round what is going on in the experiment with two holes, he also commented: 'I think I can safely say that nobody understands quantum mechanics . . . Nobody knows how it can be like that.'† So you are in good company.

The experiment with two holes is also called the double-slit experiment, because when it is carried out using light the holes can simply be two parallel slits made in a piece of card or paper with a razor. Light is shone through the two slits in a darkened room, and spreads out on the other side before arriving at a second sheet of card, where it makes a pattern. The pattern is one of alternating light and dark stripes. In the nineteenth century this was explained, or interpreted, as the result of waves spreading out from the two slits and interfering with one another, like overlapping ripples spreading out from two pebbles dropped into a still pond simultaneously. Where the waves are moving in step, they combine their strength to make a bright stripe; where the waves are moving out of step, they cancel each other and leave a dark stripe. This seemed to be definitive proof that light is a wave.

But at the beginning of the twentieth century Albert Einstein proved that light behaves like a stream of particles. In a process known as the photoelectric effect, light falling onto a metal surface knocks electrons out of the surface. The energy of the ejected electrons only has certain values, and Einstein interpreted this as the result of light arriving at the surface in the form of little particles, now called

* Richard Feynman, Robert Leighton and Matthew Sands, *The Feynman Lectures on Physics Volume III*, Addison-Wesley, Massachusetts, 1965. The term 'quantum mechanics' is essentially synonymous with 'quantum physics'.
† *The Character of Physical Law.*

photons, each with a certain energy. It was, incidentally, for this work, not either of his theories of relativity, that Einstein received the Nobel Prize.

So there are two sorts of experiment you can do with light, one which shows light behaving as a wave and one which shows light behaving as a stream of particles. Exactly the same thing happened with the investigation of electrons, but the other way round.

Towards the end of the nineteenth century, experiments directed by J. J. Thomson in Cambridge proved to everyone's satisfaction that electrons are particles. But in the 1920s, experiments carried out by several researchers, including J. J.'s son, George, showed electrons behaving as waves. J. J. Thomson got the Nobel Prize for proving that electrons are particles; George Thomson got the Nobel Prize for proving that electrons are waves. Nothing better sums up the non-commonsensical nature of the quantum world.

Today, variations on the experiment with two holes are so subtle that they can be carried out by shooting single entities, photons or electrons in different experiments, through the holes one at a time. I'll describe the results for electrons, but exactly the same kind of experiments have been carried out for photons as well. Instead of a sheet of card on the other side of the experiment, there is a detector screen like a computer monitor which records a spot of light every time an electron arrives, and allows these spots to stay and build up into a pattern as more and more electrons arrive. When researchers do this, each electron arrives as a particle and makes a single spot of light on the screen. But as hundreds and thousands of electrons are fired through the experiment one after another, the pattern that builds up on the screen is an interference pattern, the typical pattern for waves moving through the experiment.

Each electron seems not only to go through both holes at once and interfere with itself, but then to find its place in the interference pattern alongside all the electrons that have gone before and all the electrons that are still to come. Entities in the quantum world seem to know about the whole experiment, both in terms of space (the two holes) and in terms of time.

There's more. The experimenters can set up detectors to look at the two holes, and monitor which one each electron goes through. When they do this, they never see the electron going through both holes at once. They see it go through one hole or the other. And when they do this, there is no interference pattern. The spots on the screen form two blobs, one behind each hole, just as you would expect if they were made by particles. The electrons also seem to know if they are being watched or not – and the same is true for photons and all other quantum entities.

This is why Feynman said that the experiment with two holes has in it the heart of quantum physics, and that nobody knows how it can be like that. We might, indeed, just as well talk of slithy toves gyring and gimballing in the wabe as of electrons going through the experiment with two holes – except for one thing. Even though we cannot *understand* what is going on in the quantum world, the equations of quantum mechanics make it possible to *describe* what is going on, with great precision. By knowing, for example, the circumstances in which electrons seem to move like waves and the circumstances in which they seem to behave like particles, we can design computer chips. It may be crazy, but it works.

INTERPRETING THE UNIMAGINABLE

So people try to come up with images of how it works in terms that human beings can comprehend. These are called interpretations of quantum physics. The first of these aids to the imagination to be developed is called the Copenhagen Interpretation, because it was largely developed by scientists working in that city. It was the standard way of thinking about the quantum world from the 1930s to the 1980s, and is still widely taught. But it raises at least as many questions as it answers.

According to the Copenhagen Interpretation, it is meaningless to ask what atoms, electrons and other quantum entities are doing when we are not looking at them. And we can never be certain what the precise outcome of a quantum experiment will be. All we can do is calculate the probability that a particular experiment will come up with a particular result. This is exactly like the way that if you roll a pair of true dice there will be a certain probability of getting a score of 12, another probability of getting a total of 5, and so on. You also know you will never get a total of 17, or 4.3. The same sort of thing happens in quantum experiments. Some outcomes are more likely, some are less likely, and some are impossible.

With dice, you may not know what total you will get in advance, but at least you know the dice are there even if you are not looking at them. When quantum entities are not being observed, the Copenhagen Interpretation says, they dissolve into a mixture of waves (sometimes called a wave function) representing the various probabilities. This mixture is called a superposition of states. When a measurement is made, the act of measuring forces the quantum entity to choose one of these states, in line with the various probabilities, and the wave function collapses. But as soon as the measurement has been made, the quantum entity once again begins to dissolve into a mixture, a new superposition of states.

Looking specifically at the experiment with two holes, the Copenhagen Interpretation says that as soon as the electron leaves the gun on one side of the experiment it dissolves into a superposition of states, waves that pass through both holes. Once the waves have gone through the holes, they interfere with one another to produce a new superposition of states. Then, when the superposition reaches the target screen the wave function collapses into a single point and the electron becomes a real particle, at least temporarily. But if we set the experiment up to see which hole the electron goes through, the act of observation forces the wave function to collapse at one of the holes, and it then spreads out on the other side from a single site, with no interference, creating a different kind of pattern.

The situation has been summed up neatly by Heinz Pagels, in his book *The Cosmic Code*. It is worth quoting his comments, since at the time he was President of the New York Academy of Sciences, and surely knew what he was talking about. He says that according to the Copenhagen Interpretation:

There is no meaning to the objective existence of an electron at some point in space, for example at one of the two holes, independent of actual observation. The electron seems to spring into existence as a real object only when we observe it . . . reality is in part created by the observer.

You may think this is absurd. If so, you are in good company. Erwin Schrödinger, another physicist who received a Nobel Prize for his work in quantum physics, hated the Copenhagen Interpretation (he once said of the quantum theory he had helped to father, 'I don't like it, and I wish I'd never had anything to do with it'), and dreamed up his famous 'cat in a box' experiment to highlight its absurdity. This is purely an imaginary scenario – a 'thought experiment'. No cat has ever suffered the indignities Schrödinger describes. But that doesn't make it any less powerful as an indictment of the Copenhagen Interpretation.

Schrödinger's own version of the parable involved radioactive atoms being monitored by sophisticated Geiger counters. My version is slightly different, and brings out an additional feature of the weirdness of the quantum world.

THE MOTHER OF ALL QUANTUM CATS

Imagine a box, perhaps about the size of a shoe box, that is completely empty except for a single electron. The Copenhagen Interpretation says that the wave function of the electron spreads out to fill the entire box, so that if we look inside there is an equal chance of finding the electron anywhere in the box. Now imagine sliding a smooth, upright partition exactly down the middle of the box, like the dividing partitions used by magicians in the illusion of sawing a lady in half. Common sense tells us that the electron must now be trapped in one half of the box – that would certainly be the case if it was a tiny ball bouncing around inside the box. But the Copenhagen Interpretation tells us that the wave function of the electron still fills both halves of the box. This corresponds to an equal probability of finding the electron on either side of the partition, if we take a look.

Now comes the additional feature. Imagine the box divided completely into two separate halves, as with the best sawing-the-lady-in-half illusions, so the two halves can be separated, with a gap between them. The Copenhagen Interpretation still says that the wave function of the electron fills both halves of the box equally. You could take one of the half-boxes on a trip to the Moon, or farther, and this would still be the case, even though the wave function did not exist in the gap between the boxes. The wave function only collapses to become an electron located at a point when you look inside one of the half-boxes. It doesn't matter which one. If you look in box A and see an electron, the wave function disappears from box B; if you look in box A and *don't* see an electron, the wave function disappears from box A and the electron is certain to be in box B. If you do see the electron, once you stop looking the wave function spreads out again, but only to fill the half-box where we now know the electron is located.

This aspect of quantum weirdness wasn't Schrödinger's main concern when he dreamed up his so-called 'cat paradox', published in 1935. He was highlighting another feature of quantum weirdness, the superposition of states. In my variation on the theme, go back to the stage where the shoe box has been divided into two, and there is a 50:50 chance of finding the electron on either side of the partition. Imagine that the box is in a large, sealed room, where a single cat lives in quiet comfort, with plenty to eat and drink. But the box is connected to a detector, which at some appointed time will make a measurement to see if the electron is in one particular half of the box. If there is no electron there, nothing happens. But if an electron is detected, what Schrödinger called a 'diabolical device' shatters a flask of poison, which floods the chamber and kills the cat.

Or does it? Common sense says that there is a 50:50 chance that the cat will survive, and a 50:50 chance that the cat will die. The Copenhagen Interpretation says that because no outside observer has seen what is going on, when the half-box is examined instead of the electron wave function collapsing the wave function of the whole room moves into a superposition of states, one corresponding to a live cat and one corresponding to a dead cat. In Schrödinger's words, 'the wave-function of the entire system would express this by having in it the living and the dead cat (pardon the expression) mixed or smeared out in equal parts.'* The cat is both dead and alive at the same time (or, if you prefer, neither dead nor alive) and stays that way until somebody opens the door to the room and looks inside.

* Translation from Wheeler and Zurek.

Then, the wave function collapses – not at the moment the door is opened, but as if it had happened at the moment the automatic monitoring equipment looked inside the box, with all that that implies for the appearance of the possibly dead cat.

The weirdness doesn't end there. Other physicists were quick to point out that this could lead to an infinite regression. If you are the only person that looks inside the room, do you make the wave function collapse, or do you become part of the superposition of states? And if a friend phones you to ask about the outcome of the experiment, does the wave function collapse then, or does the friend become part of the superposition? Taken to its logical extreme, this line of argument raises the question, seriously debated by cosmologists, of who (or what) can observe the entire Universe and make its wave function collapse into a definite state. Why isn't everything hung up in a superposition of states?

If there was anything better than the Copenhagen Interpretation, it would have been discarded long ago. But there isn't anything better. There are only alternative interpretations,* which are precisely as good as the Copenhagen Interpretation, in the sense that they are just as good at predicting the outcomes of quantum experiments, but no better, because they do not predict anything the Copenhagen Interpretation does not predict. And they each involve inevitable components of quantum weirdness, such as signals that travel backwards in time or instantaneous communication between quantum entities across great distances. So which quantum interpretation you choose to work with is simply a preference based on which aspect of quantum weirdness you feel most comfortable (or least uncomfortable) with. The one that is relevant to the search for the Multiverse, and which is exactly as good as all the other interpretations, including the Copenhagen Interpretation, as far as any experimental test goes, is the Many Worlds Interpretation, developed by Hugh Everett in the 1950s.

* Discussed in my book *Schrödinger's Kittens.*

THE MANY WORLDS OF
HUGH EVERETT

Everett was born in Washington, DC, on 11 November 1930. He clearly had a precocious interest in the big question of life, the Universe and everything, since in his thirteenth year he wrote to Albert Einstein to ask what it is that holds the Universe together; in a letter dated 11 June 1943, Einstein replied that 'there is no such thing like an irresistible force and immovable body.'* After graduating from high school, Everett studied chemical engineering at the Catholic University of America and also in Washington, receiving his first degree in 1953. One of his friends at the University, Karen Kruse, later married the science-fiction writer Poul Anderson, who was himself a physicist and became an enthusiast for Everett's version of the Multiverse idea, which coloured several of his stories.

By the time he received his bachelor's degree, Everett's own interests had turned towards theoretical physics, but in order to take his education further he needed financial support. As an outstanding student, he was offered a prestigious National Science Foundation Fellowship to work for a PhD in mathematics at Princeton, which he was happy to accept. This was at the height of the Cold War, and the terms of the Fellowship required him to work on game theory, which in spite of its cosy name has important military applications. Everett did the work he was supposed to do, but as soon as he was safely installed at Princeton he also began looking for a way to transfer to the Physics Department. He made the transfer at the beginning of his second year at Princeton, in September 1954, and was initially assigned to the care of Frank Shoemaker as his thesis adviser. But although physics did become the subject of Everett's PhD work, he also continued his work on game theory.

Almost immediately after officially becoming a physicist, Everett came up with the big idea that he is now remembered for. Following a party at which a considerable amount of sherry was consumed,

---

* Everett-related quotes are taken from an unpublished 'Biographical Sketch' by Eugene Shikhovtsev.

Everett, his fellow student Charles Misner (who would later become a leading expert on relativity theory) and a visitor, Aage Petersen, amused themselves by dreaming up increasingly ridiculous implications of quantum puzzles like the parable of Schrödinger's cat. Their choice of subject was thanks to the presence of Petersen, who was then working as an assistant to Niels Bohr, one of the quantum pioneers and the leading proponent of the Copenhagen Interpretation. The puzzles, like Schrödinger's cat, all stem from the difficulty of understanding and interpreting what happens when the wave function collapses. Everett's big idea, initially tossed out more or less as a joke, was to ask, what if the wave function doesn't collapse? What if the superposition of states stays forever?

In the cold light of the following day, the wild idea didn't seem so wild to him after all, and Everett decided to investigate it properly, using the equations of quantum theory. But first he had other duties – a lecture on military applications of game theory that he gave in December 1954, and the small matter of graduate examinations, which he took in the spring of 1955. He received his Master's degree after passing those exams. So it wasn't until the summer of 1955 that he began to write up his big idea and its implications in proper mathematical language. The result was a draft thesis, typed up by his girlfriend Nancy Gore, who Everett married the following year. The subject was outside the scope of Frank Shoemaker's expertise, so, armed with this material, Everett transferred from Shoemaker to John Wheeler as his thesis adviser – in fact, he had already discussed the idea with Wheeler before writing the draft dissertation.

Wheeler was the perfect man for the job. He was born in 1911, and in the mid-1930s, shortly after finishing his own PhD, he had worked with Niels Bohr in Copenhagen for a couple of years. Soon after returning to America he had been Richard Feynman's adviser when Feynman was a PhD student at Princeton. An expert on the general theory of relativity, he would be the person who coined the name 'black hole' in its modern astronomical context, in 1967. Wheeler was always open to new ideas, and happy to encourage their development even if he did not always agree with them.

At the beginning of his third year at Princeton, in September 1955, Everett wrote two short papers for Wheeler developing his idea; these are now in the archive of the Niels Bohr Library of the American Institute of Physics, along with other Everett documents. In one of these papers, Everett writes for the first time about the 'splitting' of an observer whenever a quantum measurement (such as looking in the shoe box for the electron) is made. Wheeler wrote in the margin, 'Split? Better words needed.' But Everett disagreed, and used the analogy of the division, or splitting, of an 'intelligent amoeba with a good memory', even though Wheeler was not keen on it. Today, it seems an almost ideal way to explain how Everett's version of quantum physics works.

On this picture, in a situation like the electron in the shoe box, after the partition is lowered, when an observer opens the lid on one side of the partition and looks inside there is no collapse of the wave function. Both outcomes are equally likely so both are equally real. The wave function does not collapse, but the entire Universe, including the observer, splits. In one branch of reality, there is an observer who sees an electron. In the other branch of reality, there is an observer, identical to the first observer up to that point, who does not see an electron. Amoebas reproduce by splitting in two. If there were an intelligent amoeba with a good memory, before the split there would be one individual, but after the split there would be two individuals with identical memories up to that point, who would then lead separate lives developing along different paths. The difference is that in the quantum case, it isn't so much that the Universe, or the observer, splits, but that the overall wave function, the superposition of states, has built in to itself a bifurcation at the moment in time where the measurement, or observation, is made. Everett's great achievement was to express this in accurate mathematical language, and to prove that his version of quantum physics is identical in every way that can be tested to Bohr's version of quantum physics, the Copenhagen Interpretation.

Revised slightly in the light of Wheeler's comments, the 137-page dissertation typed up by Nancy was circulated to various experts, including Niels Bohr, for further comments in January 1956. Everett left Princeton in the spring of 1956 to take up a career with the Pentagon, working on highly classified material for the Weapons Systems Evaluation Group, in which he soon became head of the mathematics division. Much of his work there is still classified, but it is understood to have involved, among other things, determining the best methods for selecting targets for nuclear attack, and the development of the concept of Mutually Assured Destruction (MAD). Everett returned to Princeton in September for his final examinations, and submitted a much shorter version of his thesis, revised in response to the comments he had received and with considerable advice on presentation from Wheeler, in March 1957. The 'splitting' did not appear in the final version, Wheeler having convinced Everett that as far as getting his PhD was concerned, discretion was the better part of valour.

Everett formally completed the requirements for his PhD in April 1957, and a paper which is essentially the same as the final version of his PhD thesis was published in the journal *Reviews of Modern Physics*, under the title ' "Relative State" Formulation of Quantum Mechanics', in July that year. Hardly anyone took any notice. One of the few people who did express interest in the work was the physicist Bryce DeWitt, but even he was initially opposed to the idea that the physical world could divide repeatedly each time it was faced with a quantum choice. Persuaded by Everett that there was something more to the idea than abstract philosophizing, DeWitt was eventually instrumental in drawing the many worlds idea to the attention of a wider audience – but that wouldn't happen for more than a decade.

Everett chose the term 'relative state' to emphasize the relationship between his ideas and those of Einstein's general theory of relativity. In Einstein's theory, there is no special place in the Universe – all observers are equally entitled to their point of view. In Everett's theory, although he did not express it in quite this way, there is no special universe in the Multiverse – all quantum states are equally real. Combining this with Einstein's insight, *all observers in the Multiverse are equally entitled to their point of view*. In a paper that appeared alongside Everett's paper in *Reviews of Modern Physics*, Wheeler drew attention to this. 'Nothing quite comparable can be cited from the rest of physics,' he wrote, 'except the principle in general relativity that all regular coordinate systems are equally justified.' All observers are equally real. Although the Many World's Interpretation is not the only version of the Multiverse, there is nothing in any of the variations on the theme to refute this insight.

It's worth emphasizing this, because people still argue about what the 'relative state' formalism actually means physically, as if Everett had left it ambiguous in some way. Far from it. Quite apart from the references to splitting and intelligent amoebas that Wheeler persuaded him to leave out of his thesis, in a footnote to the *Reviews of Modern Physics* paper Everett wrote, 'From the viewpoint of the theory, *all** elements of a superposition (all "branches") are "actual", none any more "real" than the rest. It is unnecessary to suppose that all but one are somehow destroyed.' And in the long draft version† he wrote:

At this point we encounter a language difficulty. Whereas before the observation we had a single observer state afterwards there were a number of different states for the observer, all occurring in a superposition. Each of these separate states is a state for an observer, so that we can speak of the different observers described by the different states . . . In this situation we shall use the singular when we wish to emphasize that a single physical system is involved, and the plural when we wish to emphasize the different experiences for the different elements of the superposition. (e.g., 'The observer performs an observation of the quantity A, after which each of the observers of the resulting superposition has perceived an eigenvalue.')

He also had a ready response for people who asked why we don't feel any of this splitting. The *Reviews of Modern Physics* paper again:

[The] total lack of effect of one branch on another also implies that no observer will ever be aware of any 'splitting' process.

Arguments that the world picture presented by this theory is contradicted by experience, because we are unaware of any branching process, are like the criticism of the Copernican theory that the mobility of the earth as a real physical fact is incompatible with the common sense interpretation of nature because we feel no such motion. In both cases the argument falls when it is shown that the theory itself predicts that our experience will be what in fact it is. (In the Copernican case the addition of Newtonian physics was required

---

* His emphasis.
† Published in the collection of papers edited by DeWitt and Graham.

to be able to show that the earth's inhabitants would be unaware of any motion of the earth.)

But the universes would be completely cut off from one another after the split: 'There is no possible communication between the observers described by these separate states,' Everett wrote in the draft thesis.

THE BRANCHING TREE OF HISTORY

Communication between the different branches of Everett's Multi-verse, sometimes known as 'parallel worlds', would be impossible, according to the same equations that describe the existence of such multiple realities. Except for one intriguing possibility, which is strictly outside the scope of this book, but too enticing to resist mentioning briefly.

That possibility is time travel. The idea of parallel worlds is one of many which appeared in science fiction long before it became respect-able science; another is the idea of time travel. In many parallel-universe stories, either the protagonists are somehow shifted 'sideways in time' into another universe, or the entire story is set in a parallel reality, an alternative history that has branched off from our own timeline at some critical point in the past – for example, where the Axis powers won the Second World War.

The resulting image we have is of history as a tree with many branches, representing different universes that exist because they branch off at different times according to different such outcomes; the analogy is actually far from perfect, since if Everett is right there is no 'main trunk' for the tree, and the branching is more complex, but it will do. Many time-travel stories involve travellers who go back in time and, either deliberately or accidentally, change history so that they return to a 'present' very different from the one they set out from.

Combining these two ideas, a traveller might go back in time down one branch of history (in one universe), then forward in time up another branch (another universe). It isn't that he or she has changed history; both versions always existed. In the classic 'granny paradox', for example, a traveller goes back in time and accidentally causes the death of her own grandmother, before the traveller's mother has been born. In that case, if there is only one timeline, the traveller is never born, so she never goes back in time, so granny survives – and so on. In the many worlds version of the story, the traveller goes back and granny is killed, but this is the branching point for another universe. The traveller might go forward in time up the alternative branch to find a present where she had never existed, or she might go back up her original branch to find, to her surprise, that granny wasn't dead after all. Either way, there is no paradox.

It all makes for entertaining fiction. The surprise for us is that there is nothing in the known laws of physics to prevent time travel, although it would be extremely difficult to build a time machine.* The equations of the general theory of relativity (the best theory of space and time we have, which has passed every test yet devised) allow for the possibility of time travel; but they don't allow the possibility of travelling back in time before the moment that the time machine is built. This is why we are not overrun with visitors from the future. The time machine hasn't been built – yet.

But getting back to the Multiverse, the granny paradox reminds us of the parable of Schrödinger's cat. This is just the kind of puzzle that Everett's Many Worlds Interpretation resolves.

Because of the way the original cat puzzle is set up, there are only two possible outcomes – two 'eigenvalues' in the language of quantum physics. The cat is either dead or alive. According to Everett's interpretation, this means that there are two equally real worlds, superimposed on one another, but never able to influence each other – a universe with a dead cat and a universe with a live cat. It is easy to imagine a more complicated scenario, with the outcome determined by the equivalent of rolling dice, in which there might be many more possible results. Extending Schrödinger's own idea, there might be a couple of dozen cats housed in their own compartments, and which one gets killed would depend on the outcome of the roll of the dice. There would be a corresponding number of parallel universes after the experiment was carried out, with a whole variety of quantum cats to consider. But not, as is often assumed, an infinite number of parallel

---

* See Kip Thorne's book. Thorne, incidentally, was another of Wheeler's students.

universes. Although the 'many' in the Many Worlds Interpretation would be an incomprehensibly large number, there is no reason to think that it would actually be infinity. In this vast stack of parallel universes, the universe next door would be almost indistinguishable from our own, universes slightly farther away would be more different, having branched off from ours at earlier times, and universes far away across the Multiverse would be utterly different from our own.

The best reason for taking the Many Worlds Interpretation seriously is that nobody has ever found any other way to describe the entire Universe in quantum terms. Wheeler realized this from the very beginning; in his 1957 *Reviews of Modern Physics* paper, his final sentence read:

Apart from Everett's concept of relative states, no self-consistent system of ideas is at hand to explain what one shall mean by quantizing a closed system like the universe of general relativity.

In 1957, such a statement made less impact than it does today. Fifty years ago, our understanding of the Universe was much more limited than it is now – so much so that there was still a lively debate between those cosmologists who argued that the Universe as we know it had begun in a Big Bang at a definite moment in time, and those who argued that it had existed forever in a Steady State. One of the great scientific achievements of the late twentieth and early twenty-first centuries has been to establish that the Universe did begin in a Big Bang, almost exactly 13.7 billion years ago, and has been expanding ever since in line with the description of space and time provided by the general theory of relativity. The more sure cosmologists are that they understand the Universe we see around us, the more obvious it is that the only way to reconcile this view with quantum physics is to take on board the Many Worlds Interpretation – which is why Everett's ideas are now regarded as more respectable than ever.

EVERETT COMES IN FROM THE COLD

It was Bryce DeWitt who first made the specialists, and then, indirectly, the general public, aware of Everett's work. DeWitt had known about it from the start, and corresponded with Everett in 1957; indeed, the note in the *Reviews of Modern Physics* paper comparing the lack of any sense of splitting with the lack of any feeling of the earth's motion stemmed from that correspondence. DeWitt was nearly eight years older than Everett (he had been born on 8 January 1923), and had graduated from Harvard in 1943. Because of war work, he only completed his PhD (also at Harvard) in 1949, and after short spells in India and Europe (he married a French physicist, Cecile Morette) he settled down at the University of North Carolina in Chapel Hill, where he worked on the quantum theory of gravity.

In 1968, by which time he was a senior figure in the field, DeWitt received a visit from the physicist and philosopher of science Max Jammer, who was planning to write a book about the history of quantum physics and its various interpretations. To DeWitt's surprise, it turned out that Jammer had never heard of Everett, and he realized that the 1957 paper had been almost entirely forgotten (as it happens, there was a reference to the 1957 paper in a list in a footnote in an earlier book by Jammer, but he wasn't the first person to include in such a list references to papers he hadn't actually read). DeWitt decided that, since Everett was no longer working in physics, he would try to rectify the situation himself, and wrote an article on the Many Worlds Interpretation which appeared in the magazine *Physics Today* in September 1970. It was this article that made many physicists (including myself) aware of Everett's work. It refers to 'the universe as continually splitting into a multiplicity of mutually unobservable but equally real worlds', discusses Schrödinger's cat, and spells out that 'every quantum transition taking place on every star, in every galaxy, in every remote corner of the universe is splitting our local world on earth into myriads of copies of itself.'

DeWitt describes 'the shock I experienced on first encountering this multiworld concept. The idea of 10^{100+} slightly imperfect copies of oneself all constantly splitting into further copies, which ultimately become unrecognizable, is not easy to reconcile with common sense. Here is schizophrenia with a vengeance.' He uses the analogy that '[the] state vector is like a tree with an enormous number of branches,' but makes the point that 'the wave function of a finite universe must itself contain only a finite number of branches.' Everett is on record (see the article by Shikhovtsev) as saying that he 'certainly approves of the way DeWitt presented his [Everett's] theory'.

Many of these ideas were further elaborated by DeWitt's student Neill Graham in his PhD thesis, and together DeWitt and Graham edited a book, *The Many-Worlds Interpretation of Quantum Mechanics*, which appeared in 1973 and included Everett's longer version of his thesis, his and Wheeler's *Reviews of Modern Physics* papers, and the article from *Physics Today*. It was this book that popularized the term 'Many Worlds Interpretation', and in the opening paragraph of their Preface the editors refer to 'a reality composed of many worlds . . . reflecting a continual splitting of the universe'. In 1974, Jammer's book* appeared, with a substantial section on the Many World's Interpretation. 'The multiuniverse theory,' he wrote, 'is undoubtedly one of the most daring and most ambitious theories ever constructed in the history of science.'

DeWitt and Graham spread the word about Everett and the Many Worlds Interpretation to physicists; it was spread farther in December 1976, when an article appeared in the 'Science Fact' column of the SF magazine *Analog* under the title 'Quantum Physics and Reality'. The subheading read 'Alternate universes are not merely gimmicks for SF writers – they're necessary for the salvation of quantum physicists!' and authors Michael Talbot and Lloyd Biggle Jr gave a clear account of the idea.

Since more people (especially students) read *Analog* than bought *The Many-Worlds Interpretation of Quantum Mechanics*, this raised Everett's profile to its highest level. By this time, both DeWitt and Wheeler were working at the University of Texas at Austin, where in 1977 they organized a meeting to discuss the problem of consciousness and whether a computer could ever be conscious. Everett was invited to attend and was the star of the show, speaking to a packed audience;

* *The Philosophy of Quantum Mechanics*, Wiley, New York.

the 'no smoking' rule in the auditorium was suspended during his four-hour talk for his personal benefit – Everett was a chainsmoker who couldn't work without a cigarette.

This would be Everett's last 'public appearance', and the only one he made as a famous scientist; but perhaps the most significant event that occurred on that trip to Austin was his meeting with one of Wheeler's students, David Deutsch, from England. Deutsch was in effect DeWitt's student as well, and was deeply interested in the nature of the 'universes' which are described by the equations of quantum physics. The two of them discussed the issue intently over lunch one day, and Deutsch recalls that Everett was extremely enthusiastic about the term 'many universes', and not wedded to the more abstract technical term 'state vector'.

Deutsch went on to become the leading proponent of the many worlds idea. He is now at the University of Oxford, and has recently made a significant breakthrough by showing how the probabilistic rules of the quantum world arise naturally in the Many Worlds Interpretation – the way the branching gives us the illusion of probabilistic outcomes of measurements.

But if the story of Everett's version of quantum physics has a happy ending, the same cannot be said of the man himself. A former colleague, John Barry, has described him as 'brilliant, slippery, [and] untrustworthy'.* He was a cold individual who loved computers but was almost a stranger to his own family, smoked and drank to excess (he was probably an alcoholic) and had an unhealthy diet – he would argue vociferously with anyone who would listen that medical science was mistaken about cholesterol being dangerous. On 19 July 1982, medical science had the last word, when Everett was found dead in bed of a heart attack, at the age of 51. It was twenty-five years, almost to the day, since the publication of the *Reviews of Modern Physics* article. In 1996, Everett's daughter, Liz, committed suicide, and in 1998 his wife, Nancy, died of lung cancer – possibly a passive victim of Everett's smoking habit. The surviving member of the family, Hugh Everett's son Mark, found fame as a song writer and leader of the band Eels; his gloomy music provides a graphic insight into what life

* *Scientific American*, December 2007, p. 78.

was like growing up with such a dysfunctional father. We can only hope that things turned out better for all of them in some of the other universes.

Which brings me back to the one thing that makes the whole Many Worlds Interpretation stick in people's throats. It seems so extravagant in terms of universes! It makes complete sense in terms of the physics and mathematics, and it is particularly sparing in terms of assumptions – the only assumption Everett made was that the equations are telling the truth. But, as DeWitt said, 'the idea of 10^{100+} slightly imperfect copies of oneself all constantly splitting into further copies, which ultimately become unrecognizable, is not easy to reconcile with common sense.'

Let's face it – what we human beings are really interested in is how *we* got to be here, and where our own world is going; surely you only need one physical world to account for the fact that we are here? That's what people used to think. But over the past two decades it has become increasingly clear that there is something odd about our Universe – something odd which allows us to be here to ask such questions. This provides a powerful incentive to take the Multiverse seriously, irrespective of the specific merits of Everett's archetypal variation on the theme. It is quantum physics that gives a solid scientific basis to the Multiverse idea; but it is the existence of an array of cosmic coincidences that points up the need for such an idea in the first place.

2

Cosmic Coincidences Revisited

What is it that makes our Universe special? In 1989, in our book *Cosmic Coincidences*,* Martin Rees and I looked at the evidence that there is something odd about the laws of nature that make our Universe so suitable as a home for life. Rees, now both Astronomer Royal and President of the Royal Society, later developed some of these ideas in his book *Just Six Numbers*. In this chapter, I want to give just a brief résumé of these cosmic coincidences – there are actually more than six, but Rees's selection will be ample to make the point. It is these coincidences, above all else, that highlight the need to take the idea of the multiverse, in one form or another, seriously.

The investigation of these cosmic coincidences is sometimes referred to as 'Anthropic Cosmology', because our existence depends on the existence of the coincidences. This is an unfortunate choice of name, since it implies that there is something special about human beings; what really matters is that the coincidences are essential for the existence in the Universe of life, or more specifically, life as we know it. The term was coined by the theorist Brandon Carter in 1973, for his presentation to a conference to mark the 500th anniversary of the birth of Nicolaus Copernicus, and appeared in print in the Proceedings of the conference, published in 1974. That explains Carter's choice of terminology, which he has since regretted; the Copernican revolution is perceived as having displaced 'mankind' from the centre of the Universe, and Carter was pointing out that although our presence may not be central to the Universe, it is in some sense privileged. The term was popularized in 1979, in a paper in the journal *Nature* by

* Later re-issued as *The Stuff of the Universe*.

Bernard Carr and Martin Rees, in which they summed up all of the relevant coincidences they were aware of at the time. But two of the best, and most easily understood, examples of what would now be called 'anthropic reasoning' were made long before the term was coined – one by the British theorist Fred Hoyle, and the other by the American Robert Dicke.

THE CARBON COINCIDENCE

These coincidences only became apparent with the development of the modern understanding of the Universe in the second half of the twentieth century. As we have seen, there is now abundant evidence that the Universe we see around us began in a hot, dense fireball – popularly known as the Big Bang – about 13.7 billion years ago. Since then, the Universe has been expanding and cooling, but within the expanding Universe clouds of gas have been pulled together by gravity to form stars and galaxies. The material stars and galaxies – along with planets and people – are made of is embedded in a sea of dark matter, detectable only by its gravity, which helped to pull things together in this way. I shall have more to say about dark matter and the Universe at large, but it is not relevant to these two coincidences. Indeed, neither Hoyle nor Dicke was aware of the existence of dark matter when they drew attention to the coincidences.

We are made of atoms, which are themselves composed of protons and neutrons (relatively heavy particles in the central nucleus of an atom) and electrons (which in this context can be regarded as relatively light 'particles' in the outer parts of an atom). This kind of matter is often referred to as baryonic matter, to distinguish it from dark matter. From a comparison of observations of radiation left over from the Big Bang (the cosmic background radiation) and theoretical calculations we know that the baryonic matter that emerged from the Big Bang was almost entirely in the form of the two simplest atomic elements, hydrogen and helium. About 75 per cent was in the form of hydrogen, each atom of which consists of a single proton and a single electron, while 25 per cent was in the form of helium, with two protons and two neutrons in each nucleus and two electrons in the outer parts of each atom. This is confirmed by spectroscopic analysis of the light from the oldest stars, born when the Universe was young. All of the other elements have been manufactured inside stars, by the process of nuclear fusion, and scattered through space when stars age and die, to produce the raw materials for later generations of stars and, eventually, planets like the Earth. This is a slow process. Even today, those heavy elements – everything except hydrogen and helium – make up less than 2 per cent of the total amount of baryonic matter in the Universe. It is that fraction of heavy elements which allows the existence of planets and people. But most of the mass of our Solar System is in the Sun itself, still mostly in the form of hydrogen and helium.

Four of the chemical elements are particularly important for life. These are hydrogen itself, carbon, oxygen and nitrogen. And two of these, carbon and oxygen, are involved in Hoyle's anthropic insight.

The main way in which the heavy elements are built up inside stars is by combining helium nuclei with one another, step by step, to make bigger and bigger nuclei. A nucleus of carbon, for example, contains six protons and six neutrons, and is essentially three helium nuclei stuck together; oxygen nuclei contain eight protons and eight electrons, and are essentially four helium nuclei stuck together – or, more significantly, each oxygen nucleus is a carbon nucleus to which a helium nucleus has been added. Other processes convert some heavy nuclei built up in this way into nuclei of elements which are not so obviously made up from helium nuclei, but those details are not important here. What does matter is that the whole process has a bottleneck right at the very beginning.

Where, after all, does carbon come from? You'd expect it to be formed by adding helium nuclei to the nuclei of the next element down the ladder, beryllium, which is composed of four protons and four neutrons. But the nucleus of beryllium – the nucleus you would get by sticking two helium nuclei together – is extremely unstable, and falls apart almost as soon as it is formed. If two helium nuclei did happen to bump into one another inside a star and stick, there might just be time for a third helium nucleus to collide with the beryllium nucleus; but in the early 1950s it seemed far more likely that this would smash the unstable nucleus apart rather than allowing all three helium nuclei to stick together to form a carbon nucleus. If there were no carbon, there would be no oxygen and no other heavy elements. There would be no life – specifically, no carbon-based life forms like ourselves.

Fred Hoyle reasoned that because we exist, there must be a way round this nuclear fusion bottleneck. In 1954, he came up with a possible route from helium to carbon. It depended on the possibility that a carbon nucleus could exist in what is called an excited state, with an energy that 'resonates' with the energy of a beryllium nucleus plus a helium nucleus. There was no reason to expect this resonance on any other basis except for the fact that we exist. Carbon exists in our Universe, said Hoyle, so the nucleus must resonate in just the right way.

This resonance is a bit like the way different notes can be played on a single guitar string. The basic note of the string produces the lowest note, the fundamental, but it is also possible to play harmonics on the same string. Just as a single guitar string cannot play every possible note, only the harmonics of its fundamental, so a nucleus cannot exist with any amount of energy. But it can absorb energy, perhaps from the impact of a gamma ray, and become 'excited' for a short time before releasing the extra energy and falling back to its lowest level, in this case called the ground state. It only works if there is just the right amount of energy coming in to make the jump to one of the possible excited states, which are like harmonics. Similarly, if a guitar is left leaning against an amplifier and someone plays a loud note through the amplifier, if the wavelength of the note is right one or more of the guitar strings will vibrate in sympathy with the note – it will resonate.

Physicists measure energy in terms of electron Volts (eV), or multiples such as millions of electron Volts (MeV). By 1954, they already knew that the energy of a combination of a beryllium nucleus and a helium nucleus is 7.3667 MeV. Hoyle said that there must be an excited energy level of carbon that is just a little bit higher than this, so that the kinetic energy of the incoming helium nucleus would lift the total energy by the right amount to make a resonance. If it did so, instead of the incoming helium nucleus blasting the unstable beryllium nucleus apart, it would form an excited carbon nucleus, which would then radiate the extra energy away in the usual way and settle down into its ground state.

When Hoyle told the experimental physicists about this, they more or less laughed in his face. The idea of predicting the properties of an atomic nucleus on the assumption that, since we exist, so the nucleus must have the required property seemed ludicrous. But he persuaded them to carry out the necessary experiments on carbon nuclei, and when they did so they found that they have an excited energy level at 7.6549 MeV, just the right amount above the combined helium/ beryllium energy for the resonance to work. Extending the musical analogy, later calculations showed that the nuclear interaction has to be 'tuned' to an accuracy of 0.5 per cent for the resonance to work.

That isn't all. As I have mentioned, oxygen nuclei are made inside stars by adding helium nuclei to carbon nuclei. Because carbon nuclei are stable, and stay around inside stars for a long time, if this process resonated in the same way, all the carbon would swiftly be converted into oxygen. As it happens, the combined energy of a carbon nucleus and a helium nucleus is 7.1616 MeV. An oxygen nucleus has an excited energy level at 7.1187 MeV, just too low for the resonance to happen. When you add in the kinetic energy of the incoming helium nucleus, the gap is slightly bigger. In this case, resonance cannot occur – but only just.

This is a remarkable pair of coincidences. If the excited carbon energy level were just a little bit lower, there would be no carbon in the Universe, because none would have been made. If the oxygen energy level were just a little bit higher, there would be no carbon in the Universe, because it would all have been converted into oxygen. Either way, carbon-based life forms like ourselves would not exist. In Hoyle's own words:

[One] point of view is that some, if not all, of the numbers in question are fluctuations; that in other places of the universe their values would be different. My inclination is to favor [this] point of view . . . the curious placing of the levels in C^{12} and O^{16} need no longer have the appearance of astonishing accidents. It could simply be that since creatures like ourselves depend on a balance between carbon and oxygen, we can exist only in the portions of the universe where these levels happen to be correctly placed.*

But this was not the only point of view offered by Hoyle. He also said that from a different perspective it looks as if 'the laws of physics have been deliberately designed with regard to the consequences they produce inside stars,' and that the Universe seemed to him to be a 'put up job'. Invoking a Designer to explain the energy levels in carbon and oxygen nuclei seemed somewhat extreme to many physicists in 1965; but, as I shall discuss in Chapter Seven, it seems less so today.

WHY IS THE UNIVERSE SO BIG?

Dicke's coincidence is even easier to understand than Hoyle's coincidence, once you know that the heavy elements have been manufactured inside stars, but no less impressive. Actually, 'coincidence' is probably the wrong word; what he really came up with, and published in detail in 1961, is an explanation, based on the fact of our own existence, of why the Universe we see around us is so big and old.

The Universe has been around for 13.7 billion years, and the distance light can travel in one year is one light year, so the farthest we can see in principle, in any direction in space, is 13.7 billion light years. There may be more regions of the Universe beyond this horizon – perhaps infinitely more – but we cannot see them because light from those regions has not yet had time to reach us. For the sake of this argument, we can round that distance off to about 10 billion light years. In the past, this cosmic horizon was closer to us; in the future it will be farther away. So why should we be around to notice how big the visible Universe is just at the present stage of its development?

You might think that this is just one of those things – that we might have evolved sooner, or later. But the stars that cook up heavy elements like carbon take several billion years to run through their life cycles and scatter the heavy elements across space when they die. It then takes a further few billion years for intelligence to evolve on a

* *Galaxies, Nuclei, and Quasars.* The only change I would make to this would be to replace the word 'universe' with 'Multiverse'.

planet (or planets) orbiting one (or more) of the later-generation stars. Roughly speaking, it takes about 10 billion years for life forms like us to appear in the Universe and notice their surroundings.

Looking ahead, in the far future stars will burn out and there will be no life-bearing planets around. We are actually around at just about the earliest time we could have appeared in the Universe. The vast size of the visible Universe, which seems at first sight to highlight the insignificance of human beings on the cosmic scale of things, is actually an essential requirement of our existence.

NUCLEAR EFFICIENCY

The coincidences highlighted by Martin Rees are more technical than this simple example of the intimate relationship between life and the Universe, but they are worth discussing in a little detail to emphasize just how strange the Universe really is. The first of these special numbers is closely related to Hoyle's anthropic argument, and has to do with the efficiency with which nuclear processes convert matter into energy, in line with Einstein's famous equation $E = mc^2$.

There is a natural tendency for things to seek out the lowest available energy level. The most familiar example of this is water flowing downhill. The higher up something is in a gravitational field, the more energy it has – when you lift something up, the work that you put into the lifting is converted into this gravitational potential energy of the object. When the object falls, or is dropped, or when water flows downhill, gravitational energy is released, converted into energy of motion, kinetic energy. Light nuclei fuse together to make heavier nuclei because there is less energy associated with each particle (each proton or neutron) in a heavier nucleus.

Protons and neutrons are collectively known as nucleons, and there is less energy per nucleon in helium than there is in hydrogen, in carbon compared with helium, in oxygen compared with carbon, and so on all the way up to iron. Provided they can overcome the mutual repulsion caused by the fact that they have positive electric charge, nuclei will stick together and release energy in the process. What holds them together, in spite of their positive charge, is a force known as the strong nuclear force, which is more powerful than the electric force but only has a very short range; so nuclei only fuse when they get very close together, which happens under the conditions of extreme density and high temperature inside stars. But when they do get close, they grab hold of each other eagerly.

For nuclei heavier than iron, this picture no longer holds. There is more energy per nucleon, compared with iron, and these elements can only be made by forcing energy into the nuclei in huge stellar explosions, called supernovae. That is why such heavy elements are rare compared with things like oxygen and iron; but it doesn't affect what happens to lighter nuclei.

The most energy is actually released at the very first step of the chain. Although a lot of helium was made in the Big Bang, more is being manufactured today inside stars like the Sun, where hydrogen nuclei are combined, in a multi-step process, to make helium nuclei. Hydrogen nuclei are just single protons, and some of them have to be converted into neutrons along the way, but every time a helium nucleus is made from two protons and two neutrons, the mass of the helium nucleus is just 99.3 per cent of the combined mass of the two protons and two neutrons on their own. The other 0.7 per cent of the mass has been converted into energy, keeping the Sun and similar stars shining. What we may call the 'nuclear efficiency' of the process is a factor of 0.007. This is by far the most important step, as far as the conversion of mass into energy inside stars is concerned. All the rest of the processes up to and including the manufacture of iron only convert another 10 per cent of the mass involved into energy. So the efficiency of this first step in the chain is the most important factor in determining how long a star can live – once all the hydrogen is used up, other fusion processes cannot keep the star shining for long.

Even more significantly, the efficiency of this first step in the chain, a factor of 0.007, is also related to the strength of the strong force that binds nuclei together. If this number, 0.007, were a few per cent bigger or a few per cent smaller, the effect on the strong force would destroy the resonance that allows carbon to form when a short-lived beryllium nucleus is struck by a helium nucleus. The number also affects the way helium itself is made. As I mentioned, this is a multi-step process. The first step involves the formation of a nucleus containing a single proton and a single neutron, bound together by the strong force and known as deuterium, or 'heavy hydrogen'. A smaller value of the nuclear efficiency corresponds to a smaller strength for the strong force. If the nuclear efficiency were just 0.006, the strong force would be too weak to bind single protons to single neutrons, and there would be no deuterium or helium in the Universe, let alone any heavier elements.

On the other hand, if the nuclear efficiency were as big as 0.008, the strong force would have been so powerful that even two protons could bind together, without any neutrons involved, to make a curious kind of helium. All the hydrogen would have been converted into this kind of helium in the Big Bang. The implications of this are complicated, but among other things it would mean that there could be no stars like the Sun, because there would be no hydrogen to convert into helium in the way I have described. Any stars that did exist would burn out much more quickly than the stars in our Universe, perhaps not allowing time for intelligent life to evolve on any planets orbiting those stars. In any case, there would be no water, which is one of the fundamental requirements of life as we know it.

The actual mixture of elements that existed in such a hypothetical universe would depend on the exact value of the nuclear efficiency; but as Rees has summed it up, 'no carbon-based biosphere could exist if this number had been 0.006 or 0.008 rather than 0.007.' And this is just one cosmic number; there are many more that are equally finely tuned to our existence.

THE INCREDIBLE LIGHTNESS
OF GRAVITY

The strong force is, logically enough, the strongest of four forces that affect material things. There is another force which operates only on the scale of nuclei and particles, known as the weak force, and then there are the two forces familiar from everyday life, electromagnetism and gravity. Even though it is the most obvious force we experience in everyday life, gravity is by far the weakest of the four. The reason why it is so important to us is that our weight is caused by the pull of the entire Earth, almost six million billion billion kilograms of matter, acting together. It is simpler to represent such large numbers using exponential notation, in this case writing it as 6×10^{24} kg, which means a 6 followed by 24 zeroes. It takes the gravitational pull of all that mass put together to hold us down on the surface of the Earth with the weight we feel.

This can be put in perspective by comparing the strength of gravity with the strength of the electromagnetic force – or with one aspect of electromagnetism, the electric force. Both forces obey an inverse square law, which means that the force between two objects decreases as 1 divided by the square of the distance between the two objects, so we are comparing like with like. Whatever the distance between them, the electric force of repulsion between two protons is 36 powers of ten (10^{36}) times stronger than the strength of the gravitational attraction between the same two protons.

On the nuclear and atomic scales, gravity is utterly insignificant, and molecules are held together by electric forces without any complications caused by the gravitational interactions between atoms. These electric forces can, of course, produce an attraction, not just a repulsion, which is what holds electrons and nuclei together in atoms, and holds atoms together to make molecules. It is the electric forces operating between a few atoms (few compared with the number of atoms in the entire Earth) that hold an apple to a tree by its stalk, resisting the gravitational pull of all of the atoms in the Earth put together. There is a competition between electric forces trying to hold things together at this level, and gravity trying to break things apart. When the apple gets heavy enough and does fall, the gravity of the Earth wins this particular battle, but only by exerting literally all of its strength on one single apple.

This only happens because gravity differs from the electric force in one important way. Electric charge comes in two varieties, positive and negative, which cancel each other out. There is no overall electric charge on an atom, and there is no overall electric charge on the Earth. But gravity always adds up; the more atoms you have in an object, the stronger its gravitational pull. By the time you get to an object the size of our Moon, or a planet, its self-gravity is powerful enough to pull all the matter together into a spherical shape, but individual atoms retain their identity within the sphere; by the time you get to an object the size of the Sun, its self-gravity is powerful enough to crush atoms in the centre of the sphere and press nuclei close enough together for fusion to occur. It is the strength (or weakness) of gravity that determines how big a star is, and how quickly it burns its nuclear fuel.

In terms of powers of ten (exponential notation), we humans are almost exactly halfway in mass between an atom and a star. The mass of the Sun is, in round numbers, 2×10^{30} kg. The mass of an atom like carbon is roughly 10^{-26} kg, where the minus sign in the exponent means that the number is written as a decimal point followed by 25 zeros and a 1. The mass of a human adult in round numbers is abut 100 kg, or 10^2 kg. That means that a person weighs 10^{28} times more than an atom and a star weighs 10^{28} times more than a person. It would take as many people as there are atoms in your body to make the mass of the Sun.

In itself, this is not a coincidence. People are the most complex things known in the Universe, because we are made of a very large number of atoms (roughly 10^{28}) linked together in complicated ways. We are each made of about a hundred thousand billion cells operating together as a single living system. There are at least a hundred times more cells in your body than there are bright stars in the Milky Way galaxy. Atoms are simple things, molecules are more complex, cells and people even more complex. But in a star, things are simple once again. If you took 10^{28} people and put them together in one place, you wouldn't have a super-complex living system. Everything would be crushed by gravity, with all the complex structure destroyed, to make a star like the Sun, with nothing more complex than atomic nuclei in its heart.

As well as being complicated creatures, we are also close to the limit of how big an active animal can be and survive on the surface of the Earth. Because of the competition between electric forces holding things together and gravitational forces tending to break things apart, smaller bodies can survive more easily if they suffer a fall – even a human child seems to bounce up unscathed from repeated tumbles. But a large animal is likely to suffer broken limbs even by falling over, let alone in a fall from a tree or over a cliff. In order to be much larger than a human being and live on Earth, you have to be sturdy and ponderous, like an elephant, or live in the sea, like a whale, where the water offers support. Roughly speaking, the rule of thumb is that the volume of a body (and therefore its mass) is proportional to the cube of its linear size (its height), but the strength of its bones is only proportional to its cross-section, which depends on the square of the linear size. Since mass is proportional to volume, and the force of gravity pulling on a body (its weight) is proportional to its mass, as bodies get bigger the forces operating when they fall increase more than the ability of their bones to withstand a fall. Galileo understood this, back in the seventeenth century. He wrote:

Nor could Nature make trees of immeasurable size because their branches would eventually fall of their own weight, and likewise it would be impossible to fashion skeletons for men, horses or other animals which would exist and carry out their functions proportionably when such animals were increased to immense weight.

This is all summed up in the old adage, 'the bigger they come, the harder they fall'.

This puts the seemingly incredible weakness of gravity in a different perspective. Suppose gravity were a million times stronger, which would still leave it 10^{30} times weaker than the electric force. This would not be enough to affect atomic and molecular processes, so everything on the scale of atoms and molecules – in particular, chemistry – would operate the way it does in our Universe. But because of a trade-off between the inverse square law and the volume rule, the amount of matter needed to make a star would be a billion times less than in our Universe (a billion being the cube of the square root of a million) and planets would also be correspondingly smaller. Anything living on the surface of such a planet would also have to be very small, in order not to break apart when it fell over. There could not be anything as large as us, and nothing with the same sort of complexity as us.

Most important of all, as this high-gravity universe expanded away from its own Big Bang, gravity would be much more effective at pulling clouds of stuff together to make galaxies, which would be much smaller than galaxies in our own Universe, with the tiny stars

47

in them so close together that they would often experience close encounters. Instead of living for about 10 billion years, as stars like the Sun do in our Universe, these stars would live for only about 10 thousand years before they had used up all their fuel. Since the chemistry in such a universe would be no different from that in our Universe, there would be no time for evolution even to begin. Gravity *has* to be as weak as it is for us to exist. A truly cosmic coincidence.

COSMOLOGY'S COINCIDENTAL CONSTANT

Another genuinely cosmological coincidence concerns the rate at which the Universe is expanding. The equations of the general theory of relativity describe the relationship between gravity, matter, space and time. This means that they include an accurate description of the expanding Universe. The equations include room for a constant number, known as the cosmological constant we met earlier and written as the Greek letter lambda (Λ). Until recently, it seemed that this constant could be set to zero. Without the lambda term, the equations describe the way the Universe burst out of the Big Bang expanding rapidly, then slowed down as time passed and gravity held back the expansion, just as a ball thrown into the air starts out moving rapidly then slows down as gravity pulls on it. But at the end of the 1990s, observations of distant galaxies showed that relatively recently (compared with the age of the Universe) the expansion has started to speed up. The simplest and most natural way to explain this is if there is indeed a small, but non-zero, cosmological constant.

Within the equations that describe the behaviour of gravity, space, time and matter, the cosmological constant is a measure of a kind of springiness of space, like the springiness of a compressed spring – an energy possessed by empty space itself, which makes space expand. This energy is sometimes called the Λ field. Cosmologists believe that something very similar existed at the birth of the Universe, giving a much more powerful but very short-lived outward push which set the Universe expanding. The difference is that the cosmological constant is very small, and seems to have been the same in every cubic centimetre of space ever since the Big Bang.

When the Universe was young, and matter was packed together at high density, the strength of gravity completely overwhelmed the springiness of empty space, and the expansion of the Universe proceeded in almost exactly the way described by Einstein's equations without the constant. But the density of matter has gone down as the Universe expands, which means that the attractive force acting in each cubic centimetre has gone down as time has passed. Eventually, a point was reached where the force of gravity trying to slow the expansion of the Universe became smaller than the force associated with the cosmological constant trying to make the Universe expand faster. This happened a few billion years ago – more observations are needed to work out exactly when.

Like all energy, the Λ field has a mass associated with it. From observations of the way the Universe expands, cosmologists calculate that this mass is equivalent to 10^{-29} grams in every cubic centimetre of space, or the mass of four or five atoms of hydrogen in every cubic metre of space (and remember, a cubic metre is a million cubic centimetres). The Universe has just gone through the changeover from slowing down to speeding up, because at present the density of matter in the Universe is a little bit less than this (about a third as big), but roughly speaking the density of matter and the density of the Λ field are the same. This is a unique epoch in the history of the Universe, and part of the puzzle of the cosmological constant is that we should happen to be around at just this special time.

The deeper puzzle is why the energy of the Λ field is so small. The equations that describe the behaviour of particles and fields allow for the existence of a Λ field, which is associated with the rules of quantum uncertainty. The trouble is, the most natural value for the field that comes out of the equations is huge – 10^{120} times bigger than the actual Λ field. This is the kind of short-lived field that explains beautifully how the Universe started expanding, so physicists have confidence in the equations. They would be happy with a Λ field that big, or with one that is exactly zero (it is always easy to get constants in the equations to cancel exactly to zero), but it is very hard to explain, from first principles, why there should be a cosmological constant that is very small, but not exactly zero. Such things are allowed by the equations, but they are not very likely.

The answer may be related to our existence. If the cosmological constant were even a little bit bigger than it actually is the cosmic repulsion produced by the Λ field would have dominated the Universe from its early days, overwhelming gravity and preventing the collapse of clouds of stuff to make stars, galaxies, planets and people. We are only here because the cosmological constant is small. This realization led to the development of an anthropic argument to explain the observed value of the Λ field. It was originally developed by Steven Weinberg and later refined by other researchers, in particular Alex Vilenkin, from the mid 1990s onwards.

The argument says that in order to have a universe where there is a good chance of intelligent observers being around to puzzle over such questions, you need a cosmological constant small enough to dominate the expansion of the universe only after the process of galaxy formation is more or less complete. The constant could have any value less than this limit, but if there is any sense in which you can 'choose' a value from a variety of options, there is a statistical argument which says that the most likely value will not be very much smaller than this anthropic limit. The same statistics predict that if you pick a person at random out of a crowd that person is unlikely to be a dwarf. So we should not be surprised to discover that we live in a Universe where the cosmological constant is a bit, but not a lot, smaller than the maximum size it could be and allow the existence of intelligent observers. This is an example of what Vilenkin likes to call the 'principle of terrestrial mediocrity'. Sure enough, the maximum possible size for the Λ field that would allow the existence of galaxies corresponds to a density roughly ten times the density of matter in the Universe today, but the actual value of the constant corresponds to a density about three times bigger than the matter density today. And we are around just after the changeover from a slowing Universe to an accelerating Universe because that is when galaxies dominate the Universe.

If there is a choice of universes – if the Multiverse is real – then the observed value of the cosmological constant is exactly what it should be for universes within the Multiverse that are suitable homes for life forms like us. The evidence for the Multiverse is building up. But there's more.

RIPPLES IN A SMOOTH COSMIC SEA

It makes sense to discuss the next two cosmic numbers as a pair, because together they tell us that the Universe is surprisingly smooth, but contains just enough in the way of irregularities to allow us to exist. This smoothness is not obvious if you look at the night sky, or at photographs of the night sky taken with sensitive astronomical cameras. What we see is patches of light – stars and galaxies – in a sea of darkness. But appearances can be deceptive.

Astronomers cannot see galaxies move across the sky in a human lifetime, or even a thousand human lifetimes. Even though the galaxies travel at high speeds by human standards, they are simply too far away for the motion to be directly observable. But it is possible to measure the way individual galaxies are moving, and the way they are rotating, from the Doppler effect. This produces changes in the spectrum of light from an object which reveal the way the object is moving. From the way individual galaxies rotate, the way they move within groups of galaxies called clusters, and from computer simulations of the way gravity pulled matter together to make galaxies in the expanding Universe, we know very accurately how much matter there is in the Universe, and how it is distributed overall.

Rather than trying to add up the actual mass of everything in the Universe, cosmologists measure the amount of matter in terms of the average density of the Universe, which is usually denoted by the Greek letter omega, Ω. The exact value of Ω is related to how fast the Universe is expanding, and to the ultimate fate of the Universe. Leaving aside the cosmological constant for a moment, there is a competition between the expansion, making the Universe bigger, and gravity, trying to halt the expansion and make the Universe collapse. If the density is big enough, gravity wins; if the density is small, the expansion wins and the Universe expands forever. But there is a special value of the density, called the critical density, where the two are exactly in balance. In such a situation the Universe expands forever, but more and more slowly, until in effect it is hovering on the edge of collapse.

Conveniently, the way gravity works means that if the Universe ever has the critical density then it stays on this knife-edge, because although the density gets less as the Universe expands and gravity in that sense gets weaker, as it does so the expansion slows by just the right amount to maintain the balance. For convenience, cosmologists define the critical density as $\Omega = 1$. And they measure the actual density of matter in the Universe today as a fraction of the critical density. The critical density is equivalent to the presence of about five hydrogen atoms in every cubic metre of space – a number which may ring a bell.

The observations, and the computer simulations, show that the density of matter in the Universe today is a bit more than a quarter of the critical density. More precisely, Ω (matter) = 0.27, which corresponds to the equivalent of about one atom of hydrogen in every cubic metre of space. At first sight, on this evidence alone it looks as if the Universe is destined to expand forever. But there is something very odd about this number, a puzzle which cosmologists have been aware of for decades.

The balancing act between gravity and expansion only works if the value of Ω is *exactly* 1. If the Universe had emerged from the Big Bang with a density even slightly greater than the critical density at that time, gravity would have pulled things together very quickly, holding back the expansion of the Universe and making the value of Ω bigger and bigger. This would make the Universe collapse on itself in a 'Big Crunch'. If the Universe had started out from the Big Bang with a density even slightly less than the critical density at that time, the expansion would have spread things thinner and thinner, making the value of Ω smaller and smaller. Both of these are runaway processes – the deviation from $\Omega = 1$ rapidly increases as time passes.

We are now 13.7 billion years away from the Big Bang, and the expansion has been proceeding for all that time, as yet scarcely affected by the cosmological constant. In order for the Universe to have lasted that long, but not to have spread itself so thin that stars and galaxies could not form at all, the value of Ω in the first second after the beginning must have been very close indeed to 1. In fact, any deviation from 1 must have been smaller than 1 part in 10^{15} (one in a million billion). The amount by which the cosmic density differed from 1 in the beginning was a decimal point followed by 14 zeroes and a 1. Those are the odds against our being here, if the initial density was 'chosen' at random. Because the critical density is the only special density, it would be easy to imagine that there must be a law of nature which requires Ω to be exactly equal to one, but very hard to imagine that there is a law of nature which requires $\Omega = 0.27$. By the 1990s, many cosmologists were convinced that the only explanation must be that the value of Ω is indeed exactly 1, and always has been. In which case, where was the 'missing' mass needed to make up the other three quarters of the Universe?

There is one additional complication. The understanding of nuclear interactions which so successfully enables physicists to calculate that the mixture of material that emerged from the Big Bang must have been 75 per cent hydrogen and 25 per cent helium, and which equally successfully explains how the other elements were manufactured inside stars, also tells us how much nuclear (baryonic) material could have been produced in the Big Bang fireball. In terms of density, the total amount of baryonic material in the Universe – the stuff stars, planets and people are made of – cannot be more than 4 per cent of the critical density. The rest of the matter known to exist from the way galaxies move, 23 per cent of the critical density, must therefore be in the form of some kind of dark matter which has not yet been identified on Earth. The search for this dark matter is one of the most pressing endeavours of particle physics today, but all we can say at present, and all that matters as far as this book is concerned, is that it exists, that it is spread very evenly through space, and that it is the gravity of the dark matter that pulls baryonic matter (mostly hydrogen and helium) into gravitational potholes, where galaxies form like puddles in a badly maintained road.

I'm sure you can see where this is going. The observers who discovered, at the end of the 1990s, that the expansion of the Universe is speeding up were astonished. They were surprised because they were not cosmologists, and didn't know that cosmologists were already trying to find a way to make up the total density of the Universe to the critical density. But many cosmologists were delighted by the news. Another way of referring to a universe in which $\Omega = 1$ is to say that it is 'spatially flat'; in 1996, just two years before the discovery that the expansion of the Universe is accelerating, I wrote in my book *Companion to the Cosmos*, after summing up the situation more or less as I have done here, that 'if cosmologists wish to preserve the idea of a spatially flat Universe . . . they may have to reintroduce the idea of a cosmological constant.' A Λ-field with energy density equivalent to 73 per cent of the critical density is exactly what is needed to make all the pieces fit together beautifully, and this Λ-field is now sometimes referred to as 'dark energy'. But that still leaves the puzzle of why Ω should be indistinguishably close to 1, and this puzzle is best tackled by looking at what cosmologists mean when they say that the Universe is spatially flat.

The flatness of the Universe in three dimensions is equivalent to the flatness of a sheet of paper spread out on my desk top in two dimensions. The surface of the Earth is more or less a two-dimensional surface, in that sense like a flat sheet of paper, but it is curved round upon itself to make a sphere. This is said to be a closed surface, because it has no edges, and if you set off in one direction across the surface and keep going you will eventually get back to where you started. The general theory of relativity tells us that three-dimensional space can be curved in an equivalent way, and these predictions have been confirmed by observations of the way light seems to be bent as it passes near a massive object like the Sun. The bending is actually a result of light following the shortest path through curved space. In three dimensions, the equivalent of the closed surface of a sphere is a closed universe, which is also bent round on itself, with no sides and in which if you head off in one direction and keep going you will eventually get back to where you started. Such a universe resembles a very large black hole. The other possibility in two dimensions is to have a surface shaped like a saddle, or like a mountain pass, and extending off forever in all directions. This is an open surface with no edges because it is endless, in which you can keep going forever in one direction and never visit the same place twice. The equivalent in three dimensions is an open universe. If there were no cosmological constant, a closed universe would be bound one day to halt its expansion and collapse back upon itself. An open universe is destined to expand forever, and if there is a cosmological constant a closed universe may also expand forever.

The shape of three-dimensional space depends on the density of matter (or matter plus energy) within it, so the closed, flat and open universes described in this way exactly correspond to the three possible fates for the expanding Universe described earlier, with a flat universe corresponding to one with the critical density, $\Omega = 1$.

Expansion smooths out irregularities in the universe and makes it flatter. A common analogy is with the wrinkly surface of a prune. When the prune is placed in water and swells up, the wrinkles get smoothed out and the surface of the fruit is more even (this is the same as the principle behind the use of some beauty treatments for removing wrinkles from human faces). If you imagine inflating the size of the prune to the size of the Earth, the surface would be very smooth indeed, and it would not be obvious to anyone walking about on the surface that it was curved at all – just as it was not obvious to our ancestors that the Earth is round, not flat.

The best explanation for the spatial flatness of the Universe is that something rather like this happened when the Universe was born, in the first split-second of its existence. At that time, the powerful but short-lived equivalent of the present Λ-field produced a dramatic expansion in which the size of what would become the entire visible Universe doubled many, many times in a fraction of a second. For obvious reasons, this process is called inflation, and will be described in Chapter Five; at the end of inflation the Universe, still less than a second old, was a hot, expanding fireball in which space had become very nearly flat (which explains why $\Omega = 1$) and only very small irregularities were left to provide the seeds from which galaxies could grow. Although I shall discuss inflation in more detail later, one thing to bear in mind now is that although this does explain the flatness and smoothness of the Universe, the whole process could have stopped sooner, leaving a universe with larger irregularities in it, and with Ω still quite different from 1. There is still, in that sense, a choice of universes.

We can measure the lumpiness of the Universe today in terms of the way galaxies congregate in clusters. Clusters of galaxies are groups in which there may be thousands of galaxies each, like our Milky

Way, made up of hundreds of billions of stars. The galaxies are held together in a cluster by gravity, orbiting around their mutual centre of mass, and the speed with which they are moving can be measured by the Doppler effect. With this information it is straightforward to calculate how much faster the galaxies would have to be moving to break free, so that the cluster dissolved away. This would require a certain input of energy to the cluster as a whole, and the amount of energy required to do the job is a measure of how tightly the galaxies are bound together by gravity, which in turn is a measure of how much the Universe deviates from smoothness.

This energy is then compared with the total energy of the cluster – its 'rest mass energy', determined from Einstein's equation $E = mc^2$. Across the Universe, in every direction we look and in every large cluster that has been investigated, the ratio of the two energies is one to a hundred thousand. The largest deviations from the average density in the Universe today are only 10^{-5} (0.00001) of the average. On a globe the size of the Earth, that would be equivalent to having no hills higher than 60 metres. This number, 10^{-5}, is the measure of how small the ripples in the smooth cosmic sea are.

But galaxies are only the visible tracers of the dark stuff that makes up the bulk of the material Universe. They provide an outline of where the largest concentrations of dark stuff are, rather like the way the lights on a Christmas tree provide an outline of where the tree is. This means that the 'potholes' in which clusters of galaxies form are tiny deviations in the density of dark matter from the average, overdensities of just one hundred thousandth of the average density (1.00001 times the average density). One of the great triumphs of observational cosmology in the first decade of the twenty-first century has been the detection of irregularities exactly this size, one part in a hundred thousand, in the cosmic background radiation. This means that the irregularities, the seeds from which clusters of galaxies grew, were indeed imprinted long ago when the Universe was young.

But suppose the irregularities had been a different size. If the critical number were smaller than 10^{-5}, it would be harder for irregularities to grow. If it were as small as 10^{-6}, just one tenth of the actual value, it would be impossible for stars and galaxies to grow at all. On the other hand, if the number were much bigger, it would be easier for structure to form – too easy, because very massive concentrations of matter would form quickly and collapse into supermassive black holes, with no chance for galaxies to form and life to evolve. If the number were 10^{-4}, ten times larger than it is in our Universe, interesting things could still happen, and there could be huge individual galaxies, each containing as much matter as a whole cluster of galaxies in our Universe. But if the number were as big as 10^{-3}, only a hundred times bigger than in our Universe, there would only be black holes and radiation.

Once again, we are faced with a cosmic coincidence. It isn't obvious, from the laws of physics, why the Universe should have emerged from the Big Bang (after inflation) with any particular value of this cosmic number; but it has a value just in the range that allows galaxies, stars and people to exist. Smaller ripples, and there would be nothing interesting; larger ripples, and the universe would be too violent.

The list of cosmic coincidences is almost long enough for me to rest my case. But before I move on to look at the implications for the Multiverse, I want to mention just one other peculiarity, a more fundamental oddness about the Universe. Why should these ripples, and everything else I have discussed so far, be taking place in a space of three dimensions?

THREE DIMENSIONS GOOD, MORE DIMENSIONS BAD

Most people would never think of questioning the fact that the Universe exists in three dimensions of space plus one of time. That's just the way things are. But some of the most important discoveries in science come from asking why things we take for granted just happen to be that way – to take a famous example, why does an apple fall from a tree, and why does the Moon go round the Earth? This example, of the kind of question that led Isaac Newton to his insights about the nature of gravity, is apposite because the fact that space has three dimensions is intimately related to one of the most important things allowing planets like the Earth to exist in stable orbits around stars like the Sun – the inverse-square law of gravity, described by

Newton in the 1680s. According to this law, the force of gravity attracting two objects to one another is proportional to 1 divided by the square of the distance between the two objects. Einstein's general theory of relativity explains this relationship in terms of curved space, but this does not affect the basic observational fact that the law operates in this way. The general theory does not overthrow Newton's description of gravity, but includes Newton's description within itself – apples didn't start falling any differently, and the Moon didn't change its orbit, when Einstein came along.

Intriguingly, an inverse-square law is the only kind of law that allows stable orbits to exist. In our Universe, if the Earth's orbit were to shift a tiny bit either way, speeding up or slowing down as it moves around the Sun, the operation of the inverse-square law would shift it back to its present orbit, in an example of negative feedback. This is because of a trade-off between the speed of the Earth in its orbit and the force of attraction it feels from the Sun – in everyday language, a balance between centrifugal force and gravity. But in a universe where the law of gravity were an inverse cube, for example, planetary orbits would be unstable. A planet that slowed slightly and moved a little closer to its sun would feel a stronger force that pulled it inwards in a spiral of doom (gravity wins), while a planet that speeded up slightly and moved a little bit farther out from its sun would experience a weakening force that allowed it to drift away into space (centrifugal force wins). Even tiny changes, like those caused by the impact of a meteorite, would be disastrous, as a result of positive feedback. Similar things happen with other kinds of laws of gravity. The trade-off allowing stable planetary orbits only works for an inverse-square law.

Among other things, the general theory of relativity explains the fact of the inverse-square law by showing that the dimensionality of the law of gravity is always one less than the dimensionality of space. In a space of two dimensions, the force of gravity between two objects is proportional to 1 over the distance between them; in a space of four dimensions, the force of gravity between two objects is proportional to 1 over the cube of the distance between them; and so on. So planetary orbits are only stable in a space with three dimensions.

Around the time that researchers were making this discovery, in the first quarter of the twentieth century, they also discovered that the equations of electromagnetism, discovered by the Scot James Clerk Maxwell in the nineteenth century, also only work in a universe which has three dimensions of space and one of time. In our Universe, gravity is what keeps planets in their orbits, and electromagnetism is what holds atoms and molecules together to make people. In 1955, just before Hugh Everett came up with his many worlds idea, the British cosmologist Gerald Whitrow proposed that the reason we observe the Universe we live in to have three spatial dimensions is that observers can only exist in universes that have three dimensions of space (plus one of time). If life can only exist in three-dimensional space, and we are alive, it is no surprise to find ourselves in a spatially three-dimensional Universe.*

But that is not to say that universes with other dimensionality could not exist – just that they would be sterile. This idea led to the development, from the 1950s onwards, of the concept of an 'ensemble' of universes, with all possible universes existing somewhere, but with life only existing in a subset of those universes where conditions are suitable for life. This ensemble idea was developed quite separately from the quantum version of the many worlds idea, and still before the term Multiverse began to be used in its modern sense. In its basic form, it doesn't tell us *where* the other universes are; but it does offer an explanation of the cosmic coincidences.

THE LOTTERY OF LIFE

The best analogy is with a lottery, where six numbers have to be chosen to have a chance of winning the big prize. Suppose several million people each choose their own sets of six numbers, then one set of numbers is pulled out as the winner. After the event, the winning

---

* It doesn't seem possible that life could exist in a two-dimensional universe in any case, because complex structures could not form. You can't, for example, have a network of wires (like nerves) without the wires crossing and touching one another, and an object can't have a channel through it, like a digestive tract, without falling apart. As Gerald Whitrow put it, 'in three or more dimensions any number of cells can be connected with each other in pairs without intersection of the joins, but in two dimensions the maximum number of cells for which this is possible is only four.'

ticket seems special. But in a more fundamental way, there is nothing special about that set of numbers. By the very nature of the lottery, somebody has to win, and before the draw is made each ticket has an equal chance of winning. Somebody had to get lucky.

Maybe the set of cosmic numbers that allows life forms like us to exist in our Universe is like that. If there is a whole array of universes with different sets of these cosmic numbers, intelligence arises only in those few universes where the cosmic coincidences came up with a winning combination. Those universes stand out from the rest only with hindsight – because we, or our intelligent equivalents, are there to observe them. But in a more fundamental way, there is nothing special about those particular universes, they are each just the result of one of many equally likely combinations of cosmic coincidences. Maybe our Universe just happens to hold a winning ticket in the lottery of life. If so, there is no 'meaning' to the coincidences.

Rees is fond of another analogy, with a large shop full of ready-made suits. If there are enough suits in different sizes hanging on the pegs in the shop, it is no surprise if someone walks in off the street and finds a suit that is a perfect fit. It doesn't mean that the suit has been tailored specifically for them, any more than the existence of the cosmic coincidences means that the Universe has been tailored specifically for us. If there were actually an infinite number of different suits available off the peg, you could be certain of finding one that is a perfect fit. If there is an infinite number of universes, there is bound to be at least one that is just right for life. The English zoologist Carl Pantin expressed it in slightly more formal language in 1968:

The properties of the material universe are uniquely suitable for the evolution of living creatures. If we could know that our universe was only one of an indefinite number with varying properties we could perhaps invoke a solution analogous to the principle of natural selection, that only in certain universes, which happen to include ours, are the conditions suitable for life, and unless that condition is fulfilled there will be no observers to note the fact.

One cosmologist, in particular, has developed the analogy with natural selection to its logical conclusion, as I discuss in Chapter Seven. This is one of several attempts to explain where (or when) the other universes are. The cosmic coincidences give us a sound reason to think that other universes exist, even without any understanding of where those other universes might be. The Many Worlds Interpretation of quantum physics gives us one sound scientific mechanism for producing other universes, even without any 'need' for them. Put the two together, and the case that other universes exist is made far more than twice as strong. The other mechanisms for producing other universes will be described in the rest of this book; but before I move on to them, I want to take the Many Worlds Interpretation to its logical conclusion, which gives us a new way of thinking about both space and time, even without considering variation in the 'constants' of physics. Along the way, we will encounter what could soon be the greatest triumph of many worlds theory, and the proof of the existence of the Multiverse – the quantum computer.

3

Quantum Bits and Time Slips

Could the existence of the other worlds described by Hugh Everett directly affect our world? The usual answer is 'no'. But the leading proponent of the Many Worlds Interpretation today, David Deutsch, says that it happens all the time. What's more, he says that it is possible in principle to build an intelligent (or, at least, self-aware) computer that would be able to feel the effects of a few of the many worlds interacting with one another, and report that feeling back to us.

Deutsch is completely convinced of the reality of the Multiverse, and takes the Many Worlds Interpretation entirely at face value. He accepts that there is, for example, a vast array of universes with different versions of himself in them, so that in some he is (not 'might be', really *is*) a Professor in Cambridge instead of working in Oxford, while in others he is not a scientist at all. On a larger scale, there are many science fiction stories with 'alternative histories', in which, say, the dinosaurs never died out but developed human-level intelligence and a civilization to match our own. Deutsch says that things like this are not merely fiction – universes in which dinosaurs have developed cities, spaceprobes and computers using silicon-based microchips 'undoubtedly' exist, because 'that's what the laws of physics tell us'.* 'It can't be that there are multiple universes at the level of atoms but only a single universe at the level of cats.'

There is, though, one important difference between Deutsch's vision of the Multiverse and the version proposed by Everett. In the original,

---

* See, for example, Julian Brown, *Minds, Machines and the Multiverse*, which includes interviews with Deutsch.

and now traditional, language of the Many Worlds Interpretation, a universe is described as 'splitting' whenever it is faced with a choice of quantum possibilities, giving rise to the image of a many-branched tree and the unfortunate hidden implication that there might be a 'main trunk' from which the branches stem. It also raises the disturbing puzzle of how an electron going through one hole or the other in a double-slit experiment here on Earth can be responsible for the entire Universe, including every distant quasar, splitting in two. But the image Deutsch prefers is of a vast array of universes which all start out the same, and have identical histories up to the point where the quantum choice is made. In one universe, the electron goes through hole A, in the other universe it goes through hole B, and thereafter the histories are different, but nothing ever splits. It's like having an infinite library full of copies of books that all start out the same way on page one, but in which the story in each book deviates more and more from the versions in the other books the farther into the book you read. As long as you are comfortable with the idea of infinity – which you have to be to take any of these ideas on board – this is a much more satisfactory image than one of a library which starts out with a single book that splits repeatedly into more and more different books as you try to read it. In a further twist, Deutsch points out that the Many Worlds Interpretation allows universes to merge back together, using the old language, or to become identical again, using the language that Deutsch prefers, as if two of the books in the library have the same happy ending arrived at by different routes. This all stems from his different take on what happens during the process of quantum interference.

BEING IN TWO MINDS

In the most basic version of the quantum two-slit experiment, photons are fired one at a time through the two holes. Everett's interpretation is that this causes the universe to split into two groups of universes, in one of which a particular photon goes through one hole and in the other of which it goes through the other hole. But, says Deutsch, whichever hole the photon goes through, it ends up in the same part of the interference pattern on the other side of the experiment, as if the two groups of universes have fused back into one. The 'splitting' only concerns what is going on inside the experiment, not the whole array of universes at large. It's only if the 'choice' of which slit to go through triggers some further event (like the death or survival of a cat) that the two groups of universes diverge and go their separate ways.

In the kind of language Deutsch prefers, it's as if two books told identical stories about the adventures of some hero until the point where he has to cross a river either by fording it or by using a bridge. In one book, he goes over the bridge; in the other, he fords the river. Then, the two stories are identical once again. Again, it's only if the 'choice' triggers some further event (like the death of the hero if he falls over while fording the river and drowns) that the two groups of universes – the two stories – diverge and go their separate ways.

Deutsch sees interference like this. As he spells out in his book *The Fabric of Reality*, referring to a version of the experiment with several holes, not just two, when a single photon enters the experiment:

it passes through one of the slits, and then something interferes with it, deflecting it in a way that depends on what other slits are open;

the interfering entities have passed through some of the other slits;

the interfering entities behave exactly like photons . . .

. . . except that they cannot be seen.

Working on the principle that if it looks like a duck, quacks like a duck and lays eggs then it *is* a duck, Deutsch says that these 'interfering entities' *are* photons – photons in the parallel realities of the Multiverse. He concludes that:

Every subatomic particle has counterparts in other universes, and is interfered with only by those counterparts. It is not directly affected by any other particles in those universes. Therefore [quantum] interference is observed only in special situations where the paths of a particle and its shadow counterparts separate and then reconverge . . . interference is strong enough to be detected only between universes that are very alike.

In an effort to find a definitive way to prove that this interpretation of what is going on is correct, Deutsch came up with the idea of an artificial brain that could remember being affected by a quantum experiment. It would literally be in two minds while a photon or an electron was passing through the equivalent of the experiment with two holes.

Deutsch talks about the possibility of constructing an artificial brain which has a memory operating at the quantum level, so that it can experience quantum phenomena directly. This computer could not only keep a record of experiments that it was carrying out on itself, but could also report how it felt at any moment during the course of the experiment – it could simply print out a running commentary on what was going on. Given the present rate of progress in computation, such a computer could probably be built well before the end of the present century. The experiments it would carry out on itself would be a little more subtle than the basic two-slit experiment, but they would be equivalent to the computer having a circuit within itself that incorporated single electrons or photons going one after the other through two holes, and the intelligence feeling the effects. Deutsch expects that such an intelligence would feel the effects of the particle going both ways at once, as the intelligence split into two copies of itself (using Everett's description) and then merged back into one.

But this would be literally indescribable to human observers, who would feel no such thing. As the split occurred, there would be two copies of the observers, as well as two copies of the machine intelligence. So during the course of the experiment, the machine will be asked by the human operators how many of the two possible paths of the electron or photon it is experiencing. The key point of Deutsch's insight is that the computer will reply, 'I am observing one and only one of the two possibilities,' but it will be giving the same answer in two different universes. Then, at the end of the experiment, when both universes have merged into one, the observers, both human and machine, will see an interference pattern, proving that the Everett Many Worlds Interpretation is correct. If the Many Worlds Interpretation is wrong there will be no interference.

It is crucially important, though, that the observers do not ask the computer *which* of the two internal paths the electron or photon is following. As soon as it reveals that, the split between the worlds becomes permanent, and there will be no interference, just as there is no interference if the experimenter looks to see which slit the electron goes through in the classic double-slit experiment. This is equivalent to the collapse of the wave function on the old Copenhagen Interpretation way of looking at things. On the Copenhagen Interpretation, the computer would either report that it was feeling *both* possibilities at once, and there would be interference, or it would report that it was feeling one possibility and there would be no interference. It could never report that it was feeling only one possibility *and* produce interference.

Even in the Multiverse, though, the computer intelligence will never remember the experience of splitting in two. It is a necessary consequence of the way the experiment works, says Deutsch, that it must 'wipe out the memory of which one of those two possibilities [it] observed', because of the interference.* If it really is intelligent, the machine can infer that it must have been split, by analysing the records of the experiment; but it could not remember what it felt like to be in two minds.

Deutsch himself has become deeply interested in the mechanics of building practical computers that would operate on quantum principles, because those operating principles reveal deep truths about the nature of the Multiverse. As early as 1977, Deutsch came up with the idea of what would now be called a quantum computer, although at the time all he was interested in was the possibility of building a real, physical machine to test the many worlds idea. He is less interested in practical applications of such machines; but it turns out that there are such applications, important enough for serious amounts of money to be spent on the quest for a working quantum computer. Since the early 1980s Deutsch has made major contributions to the theory of quantum computation, which has now become a practical reality, albeit with machines far too simple to carry out this kind of test of the Many Worlds Interpretation. The fact that even these far simpler machines work at all, though, is, he argues, convincing proof of the reality of the Multiverse.

---

* See *The Ghost in the Atom*.

IN SEARCH OF THE
QUANTUM COMPUTER

Conventional computers – often referred to as 'classical' computers – store and manipulate information consisting of binary digits, or bits. These are like ordinary switches that can be in one of two positions, on or off, up or down. The state of a switch is represented by the numbers 0 and 1, and all of the activity of a computer involves changing the settings on those switches in the appropriate way. My own computer is, while I am writing these sentences using a word processing program, also playing music, showing an image in one corner from a live connection to the camera in my son's computer in his shop in Brighton, and has an email program running in the background that will alert me if a new message comes in. All of this, and all the other things computers can do, is happening because strings of 0s and 1s are being moved and manipulated inside the 'brain' of the computer.

Eight bits like this make a byte, and because we are counting in base 2 rather than base 10 the natural steps up the ladder of multiplication do not go 10, 100, 1000 and so on but 2, 4, 8, 16 and so on. It happens that 2^{10} is 1,024, which is close to 1,000, and we are used to using base 10, so 1,024 bytes is called a kilobyte. Similarly, 1,024 kilobytes make a Megabyte, and 1,024 Megabytes make a Gigabyte. The hard drive of my laptop computer can store 160 Gigabytes of information, and the 'brain' – the processor – can manipulate up to two Gigabytes at a time, all in the form of strings of 0s and 1s.

But a quantum computer is something else. In the quantum world, entities such as electrons can be in a superposition of states. This means that quantum switches can be in both states, on *and* off at the same time, in a superposition of states, like Schrödinger's cat. Electrons themselves, for example, have a property called spin, which is not quite the same as what we mean by spin in our everyday world, but can be thought of as meaning that the electron is pointing either up or down. If we say that 'up' corresponds to 0 and 'down' corresponds to 1, we have a binary quantum switch. Under the right circumstances, it can exist in a situation where it is pointing both up and down at the same time.

A single quantum switch that is in a superposition of states can 'store' the numbers 0 and 1 simultaneously. By extension from the language of classical computers, such a quantum switch is called a qubit, short for quantum bit and pronounced 'cubit', like the Biblical unit of length. The existence of qubits has mind-blowing implications. Two classical bits, for example, can represent any of the four numbers from 0 to 3, because they can exist in any of four combinations: 00, 01, 10, and 11. To represent all four of the numbers (0, 1, 2 and 3) simultaneously, you would need four pairs of bits – in effect, one byte. But just two qubits can represent all four of these numbers simultaneously. A set of bits (or qubits) operating as a number store in this way is called a register. A register made up of eight qubits (a single qubyte) can represent not four but 2^8 numbers *simultaneously*. That's 256 numbers stored in a single qubyte. Or as Deutsch would put it, it represents 256 different universes in the Multiverse, sharing the information in some way.

In a functioning quantum computer, any manipulation involving an operation on each of those 256 numbers represented by that qubyte of information is carried out simultaneously in all 256 universes, as if we had 256 separate classical computers each working on one aspect of the problem in our universe, or one computer that had to be run 256 times, once for each value of the number. Looking farther into the future, a quantum computer based on a 30-qubit processor would have the equivalent computing power of a conventional machine running at 10 teraflops (trillions of floating-point operations per second); conventional desktop computers today run at speeds measured in gigaflops (billions of floating-point operations per second), ten thousand times slower than a 30-qubit quantum computer. This hints at the prodigious power of a quantum computer; but the trick is to find a way of getting useful information out at the end of the calculation – getting the different universes to interfere in the right way to produce an 'answer' that we can understand, without destroying the useful information along the way.

It was Deutsch himself who first highlighted the power of a computer based on 'quantum parallelism', in a scientific paper published in 1985. But, as he realized, there are severe limitations on the application of such a computer, because the inhabitants of one universe cannot directly observe the results of the calculations carried out in all the parallel universes. It is only when the quantum computations interfere with one another that a limited amount of information is available to observers in all the universes involved. In an extreme example, reminiscent of the behaviour of the computer Deep Thought in *The Hitch Hiker's Guide to the Galaxy*, you might run the quantum computer and get an answer out without knowing the question that had been asked, because it had been asked by an operator (or operators) in the other universes.

So even if you had a working quantum computer, you couldn't use it simply to do the kind of things a conventional computer does, but better and faster. It wouldn't be a better word processor, for example. The only practical reason for investing large amounts of money and effort in building a quantum computer would be if there were some powerful application with practical benefits that a quantum computer could do but a conventional computer could not. For nearly ten years, the quantum computer was a solution looking for a problem to solve. Then in 1994 Peter Shor, of the Bell Laboratories in New Jersey, came up with just such a 'killer' application.

THE KILLER APPLICATION

The 'killer app' seems at first sight to be a rather mundane, even tedious, bit of mathematics. But it is its very tediousness that makes it so important. In the political, military and commercial worlds, keeping secrets is of the utmost importance, and secure codes that can be used to convey sensitive information from one person to another without third parties getting hold of it are both essential and valuable. The ability to crack such codes is, if anything, even more valuable, and has stimulated a major, though largely unsung, research effort. The best codes today, regarded as virtually uncrackable, are based on the difficulty of finding the factors of any very large number.

Factorization is something we all learned in school, though many of us have forgotten it. It is related to the multiplication of prime numbers. Leaving aside the number 1, which is a special case, a prime number is any number which can only be divided exactly by itself and 1. So 2 is the first prime number (and the only even one, since all other even numbers divide by 2), 3 is the second, 5 the third, and so on. If we multiply two prime numbers together, we get a number that is not a prime, and those two primes are the factors of that number. For example, 3 × 5 = 15, and 3 and 5 are the factor of 15. A number may have more than two factors, but there is no need for such complications in this example. Multiplying numbers is easy. But the only way to find the factors of a number is by trial and error – easy enough for a number like 15, but much harder when you get to very large numbers. The only real guide you have to go on is that one of the factors must be less than the square root of the very large number.

The reason this is relevant to cryptography is that a code can be based on a very large number made by multiplying two large prime numbers together. These are the best codes yet devised, and very widely used. The very large number is used by the sender of coded messages in scrambling the messages up, but the factors are used by the receiver to unscramble the message.* Anyone trying to crack the code would be able to find the very large number easily, but it would take a long time to find its factors, even using the best conventional computers.

Deutsch gives an example of what would be a fairly simple such problem to solve using a computer, but an impossible one for an individual human mathematician. The number 10,949,769,651,859 has just two factors, 4,220,851 and 2,594,209. How long do you think it would take you to find these factors, starting out with 10,949,769,651,859 and dividing it first by 3, then by 5, then by 7, then by 11 ... and so on until you got to 2,594,209? A computer could do it in less than a second. But 10,949,769,651,859 only has 14 digits, and each time we add a digit to the length of the number the size of the square root is about three times bigger. This is because adding another digit is roughly the same as multiplying by 10, and the square root of 10 is a bit more than 3. So adding another digit

* This is a great oversimplfication; if you want to know the details of the cryptography, see George Johnson's *A Shortcut Through Time*.

triples the time taken to find the smaller of the two primes. For a number 25 digits long, the factorization process would have taken centuries using the best computer available in 1997, when Deutsch offered this example; as computer speed increases, all you have to do to keep your messages secret is to use larger and larger numbers.

But Shor put together a package of ideas, together known as Shor's algorithm, which show how a quantum computer could factorize very large numbers more easily than a conventional computer could deal with 14-digit numbers like 10,949,769,651,859. In essence, the quantum computer carries out every possible division in an array of parallel universes simultaneously, so the answer emerges in the time it takes to do one division. Thanks to interference, Shor found a way for the answer to appear on all the computers in all the parallel worlds. Only the calculations that lead to the right answer add together in the interference process, while all the wrong answers cancel each other out.

In theory, there is still a big problem with such a computer, because it is very prone to another kind of interference, interference from the outside world, which acts like noise in the system. This means that you can't trust the answers you get out. But for this kind of problem that doesn't matter! Suppose that the quantum computer is so sensitive to this noise that it may produce the right answer only once in every thousand runs. So what? Each run only takes a fraction of a second. Just run the machine 1,000 times, or more, and use a conventional computer to multiply up the various 'answers' to find which ones really are the factors of the very large number being analysed. To take an extreme example, finding the factors of a number a thousand digits long would take a conventional computer many millions of billions of years – far longer than the time since the Big Bang. It would take a quantum computer using Shor's algorithm about 20 minutes.

These ideas were soon refined (by Deutsch and his colleagues, among others), and other algorithms that could be used to solve specific problems on quantum computers were discovered. With the genuine benefits of even a single-use quantum computer clear, by the end of the twentieth century the race was on to make a working machine that could factorize numbers in this way – starting small, of course, but with large numbers in mind.

PRACTICALITIES

The big problem with all quantum computers is getting access to the information. The computer may operate perfectly satisfactorily at the quantum level, but in order to find out what is going on we have to interfere with it from the outside – or rather, we have to allow the quantum system to interfere with the outside environment. Any interaction with the outside world disturbs the quantum processes. The quantum interference effects get spread out among larger groups of particles in the outside environment, in a phenomenon known as decoherence. Physicists usually say that it is the interference from outside that causes quantum systems to decohere; but, as Deutsch points out, this is the wrong way of looking at things, and it is really the effect of the quantum process on the outside world that causes decoherence, as quantum information is spread out more widely and lost in the noise of everything else that is going on. Either way, it means that the quantum systems have to be carefully shielded from any outside influences while a calculation is going on, then poked in just the right way to read off the answer to the calculation. Even then, the computer can only give you the answer once, since the act of reading the information itself corrupts it. Unlike a conventional computer, it cannot keep the answer stored in its memory to be read off again and again.

If such problems could be overcome, a quantum computer could operate using single electrons, each trapped inside a cage of atoms, as the 'switches' in the computer. Such a 'quantum dot' could be made to switch from one quantum state to another by tickling it with a pulse of pure light from a laser beam – or, of course, it could exist in a superposition of states. Apart from the problems of working on such a small scale, though, such a computer would be very prone to errors caused by decoherence, since each qubit would be represented by a single electron that could be corrupted by any outside influence.

The best way around this problem is to spread the information about the state of a single qubit across several (or many) different switches, so that if one gets corrupted you still have the others. An early success along these lines came in 1998, when a team headed by

Isaac Chuang, from the Los Alamos National Laboratory, and Neil Gershenfeld, from the Massachusetts Institute of Technology, found a way to spread a single qubit across three quantum states (technically, nuclear spin states) in each molecule in a liquid solution. The 'computer' is programmed using a series of radio-frequency electromagnetic pulses, and the resulting average state of all the molecules, representing the output of the program, can be monitored using magnetic fields, in an adaptation of the technique of nuclear magnetic resonance, or NMR. This technique is used in medicine to 'see' inside the body; there, it goes by the name magnetic resonance imaging, or MRI, because many patients are frightened by the word 'nuclear'.

What is particularly clever about the work of the Los Alamos/MIT team is that although measuring the state of an individual quantum dot to find out whether it has been corrupted would destroy the information it contains, the NMR approach can be used to compare the average states of the molecules to see if they are different from one another without actually measuring them. So when differences show up where there shouldn't be any, the team know that an error has occurred and can take steps to correct it.

What will probably go down in history as *the* landmark event in quantum computing – the equivalent in aeronautics would be the Wright brothers' first flight – came in 2001. A team at IBM's Almaden Research Center actually factored the number 15 using Shor's algorithm, running on a quantum computer. The heart of the computer was a molecule which contains five fluorine atoms and two carbon atoms, giving them seven nuclear spins to play with – seven qubits. But not just one molecule – they used a solution containing an estimated billion billion (10^{18}) of these molecules in about a thimbleful of liquid, monitored by NMR with the averaging over all those molecules effectively compensating for any errors introduced by decoherence. In effect, each molecule is a seven-qubit computer (equivalent to a 128-bit conventional computer), each working out the same problem and comparing their answers to check for errors.

Seven qubits is the minimum number required to factor the number 15 using Shor's technique, and the computer correctly found the factors to be 3 and 5. This proved that quantum computing works, proved that Shor's algorithm works, and makes it very difficult to doubt the existence of the Multiverse. Looking at what happened to aviation in the century following the Wright brothers' success gives the merest hint of where quantum computing might be by the year 2100.

There are other molecules that could be used in such a liquid computer, and Gershenfeld likes to point out that caffeine is a good candidate. It might well be possible, in another echo of the *Hitch Hiker's Guide*, to build a quantum computer which uses a cup of coffee as its processor. But there are, at least at present, severe limitations on the development of the liquid computer technique.

Unfortunately, the strength of the radio-frequency signal used to monitor the molecules falls off rapidly as the number of nuclei involved increases, and with present technology the largest quantum computer that could be constructed using this technique would only have ten qubits. But there are other techniques already being developed, one of which reached the landmark of the first qubyte computer, with 8 qubits, in 2005. We can no more guess how the future of quantum computing will develop than the Wright brothers could have designed the stealth bomber, and in any case such speculation has no place in this book. What matters here is *where* the calculations performed by quantum computers take place.

WHERE DOES IT ALL HAPPEN?

Many people involved in quantum computation simply don't worry about where the calculations are being carried out, as long as the computers work. Most of the ones who do think about such things have a background in mathematics, where they are used to thinking of things in terms of imaginary spaces, fictional constructs which make certain calculations easier. Unfortunately, this can make them blind to what is really going on in a quantum computer.

In the real world, the position of an object in three-dimensional space can be represented by three numbers, coordinates corresponding to its distance from some chosen reference point – usually in terms of the distance along each of three axes at right angles to one another, the three-dimensional equivalent of a two-dimensional graph. The velocity of an object is also a three-dimensional property, because it involves direction, and mathematicians are happy to think in terms of an imaginary 'velocity space' where a single point in that space represents the components of velocity in three directions at right angles to one another, which add up to give the actual velocity. You could imagine actually making a three-dimensional model of velocity space to represent this, rather like the way a three-dimensional relief globe could represent the geography of the Earth.

But why, if you are a mathematician, stop there? Why not imagine combining both sets of information in one single *six*-dimensional space, where a single point contains information about *both* the position *and* the velocity of a single particle? Even mathematicians may find it hard to picture six directions all at right angles to one another, and nobody is going to build a six-dimensional model to represent this space. But as far as the calculations are concerned the equations are exactly like those of Pythagoras' famous theorem for right-angle triangles, with a few extra terms corresponding to the extra dimensions (one for each extra dimension). For historical reasons, such an imaginary space is called 'phase space', but the term 'phase' has no significance in this context now except as a name.

Mathematicians can extend the concept of phase space more or less indefinitely. It has many applications, but the velocity/position example is good enough for now. It takes a six-dimensional phase space to describe the state of a single particle, but if there are two particles rattling around in an otherwise empty box it takes a 12-dimensional phase space to describe the situation at any moment in time. This particular kind of phase space always has six times as many dimensions as the number of particles in the box. Statistical techniques can be used to describe the way such a system changes in phase space as time passes, and these have important implications, for example in chaos theory.* But the only relevant point here is that mathematicians are used to the idea of dealing with imaginary phase spaces with absolutely enormous numbers of (mythical) dimensions. So when someone tells them that, say, a quantum computer that is factorizing a 250-digit number does so by operating in a superposition of 10^{500}

* See *Deep Simplicity*.

states,* their immediate reaction is to think of this in the same way that they think about phase space.

But there is an essential difference. The phase space representing the behaviour of particles rattling around in a box, or some other physical system, is an imaginary representation of the real, physical thing – in this case, the box full of particles. But the quantum computation is itself a real, physical thing, not something imagined by mathematicians. The calculation involves, in this case, 10^{500} real computers working together. Where are they? Deutsch makes the point forcefully:

To those who still cling to a single-universe world-view, I issue this challenge: *explain how Shor's algorithm works* ... When Shor's algorithm has factorized a number, using 10^{500} or so times the computational resources that can be seen to be present, where was the number factorized? There are only about 10^{80} atoms in the entire visible universe, an utterly minuscule number compared with 10^{500}. So if the visible universe were the extent of physical reality, physical reality would not even remotely contain the resources required to factorize such a large number. Who did factorize it, then? How, and where, was the computation performed?

You might also ask *why* was the computation performed? Why would the inhabitants of 10^{500} other universes allow us to run the program on their computers? What's in it for them?

The simple answer is that what's in it for them is the same as what's in it for us. Remember the analogy with a library of books that start out in identical fashion. All of the universes in which the computation is taking place are identical to our universe up to the point where the factorization program is set running. For all intents and purposes, the inhabitants of those particular other universes *are* us, and they run the program for the same reasons we do. During the computational process, the universes differentiate – the stories written on those pages of the books are different from one another. But after the computation, the universes are essentially identical once again.

There are, of course, vastly more universes in which there are people

---

* This is a genuine example borrowed from Deutsch; 10^{500} is, of course, a 1 followed by 500 zeroes.

who do not build quantum computers or who choose to run different programs on their computers. But those universes are so different from ours that they do not interfere, in the quantum sense, with our universe. It's only universes sufficiently similar to ours for their inhabitants to want to solve the same puzzles that we want to solve that interfere in just the right way to solve those puzzles.

This is as far as I want, or need, to take the story of quantum computation. To my mind, the fact that quantum computation works proves the existence of the Multiverse. The library book analogy, though, leads naturally to another aspect of the Multiverse which runs counter to 'common sense' – the nature of time.

A METAPHOR FOR THE MULTIVERSE

The first time you hear of the idea of parallel universes, it is natural to think of them as lying side by side, with universes that are very similar to our own close by ('next door'), and universes that are different from our own becoming increasingly different as you 'move' (whatever that means) farther away from our own time line, in a sense 'sideways' in time. This is the basis of many science fiction stories, and it also appears in some popular accounts of the Multiverse. But there is no reason to think that the Multiverse is like that. The term parallel universes is inaccurate and misleading, but unfortunately it has become part of the language of Multiverse theory; it is better to think of it as a metaphor, in the way we might say that two people who have never met but follow similar career paths lead parallel lives. A better metaphor would be to say that all the various universes are at right angles to one another, like all the directions in the mathematicians' imaginary phase space. That isn't an accurate description of the Multiverse either; but at least it has the merit of describing something totally outside our common sense experience, so that we can appreciate that the Multiverse is actually like nothing we have experienced directly, not a series of cosy 'worlds next door' like a row of nearly identical terraced houses.

Think of the Multiverse library. A librarian with a tidy mind might indeed put all the books in order, so that the ones with nearly identical stories are next to each other on the shelves, and the farther along the shelves you look the more the stories diverge. But there is no need for this. The books could be arranged in any order, or simply piled up in a heap on the floor, and they would still serve as a metaphor for the Multiverse. It would still be true that all the books start out identical to one another, and that the stories they tell begin to differentiate as you turn the pages. It would still be true that in 10^{500} or so of the books the story would be identical up to a certain point, then there would be 10^{500} or so different accounts of the factorization of a 250-digit number, then the result of the calculation would be recorded and the stories would become identical again. All that matters is that you read the pages of each book in the right order, not that the books themselves be read in any particular order.

It's easy to read the pages of a book in the right order, because they are numbered. Even if the binding gets damaged and a few pages become loose, you can soon put them back in the right place. In fact, if the pages weren't bound together at all, but were in a heap on the floor, you could still read them in the right order. If all the books in the Multiverse library actually consisted of heaps of loose pages, each one labelled with one number corresponding to the book it belonged in and another number corresponding to its place in that book, the whole heap could be jumbled up as much as you like, and the stories could still be read. They would be just as real. Even better, suppose that each page has an introductory paragraph providing a précis of the story so far. This summary would have fairly accurate information about what was in the immediately preceding pages, and more vague information about the details of the story on much earlier pages. You could pick up any page from the heap at random, see which 'history' it belonged in, find out the outline of the story so far, and get a detailed account of what was going on at that moment of the story.

If each page in each book corresponds to a moment in time, that is exactly like our everyday experience of our own universe. We have records of the past, both physical records and our memories, which tell us in detail about the recent past and in rather less detail about the distant past, and we have a clear and detailed 'view' of what is going on now. We have the *impression* of time flowing 'like an ever-rolling stream'; but all we are aware of is the present moment and our memories – our précis – of the past. Is this really so different from our experience of reading a gripping story, living each moment with the characters as we turn the pages? I believe that this means that the image of jumbled heaps of numbered pieces of paper on the library floor is not just a mere metaphor, but provides genuine insight into the nature of the Multiverse.

I confess I did not come up with the idea all by myself. I first came across something like this in 1966, when I read Fred Hoyle's science fiction novel *October the First is Too Late*. In the book, Hoyle's characters encounter a jumbling up of time in which different places on Earth experience different times simultaneously. As one of the scientists in the story explains, 'In Hawaii it is the middle of August 1966, in Britain it is 19 September 1966, on the American mainland I would guess it is somewhere before the year 1750, in France it is the end of September 1917.' Hoyle states in a note to the reader of the book that although it is a work of fiction, 'the discussions of the significance of time and of the meaning of consciousness are intended to be quite serious'; I was lucky enough to be able to confirm that he meant this when I became a student at the Institute of Theoretical Astronomy in Cambridge, where Hoyle was the Director, and asked him about the book. So what does he have to say in it about the nature of time?

WHEN DOES IT ALL HAPPEN?

The scientist in Hoyle's novel, speaking with Hoyle's words and offering what Hoyle confirmed to me was his view as a physicist of the nature of reality, first describes the world in terms of the spacetime of relativity theory. 'If you consider the motion of the Earth around the Sun,' he says, 'it is a spiral in four dimensional space-time. There's absolutely no question of singling out a special point on the spiral and saying that particular point is the present.' This is often known as the 'block universe' model of reality, which says that all moments in time exist in spacetime just as all positions in space exist – that the year 1452, or 3173, is just as real as today even though we are not experiencing it, in the same way that New York and Mumbai are just as real as London, even if you are living in London. The block universe is actually the only interpretation of relativity theory that makes sense, although many relativists prefer to ignore the fact.

Then, Hoyle makes an analogy with a stack of pigeon holes – little boxes like the ones used in post-office sorting rooms before mechanization. An infinite array of little boxes, in a numbered sequence. In each of the pigeon holes there will be information, perhaps printed on pieces of paper, or stored in a computer, describing the contents of other pigeon holes in the sequence. The information about the contents of pigeon holes with numbers lower in the sequence turns out to be pretty accurate, but the information about pigeon holes with numbers higher in the sequence turns out to be vague and sometimes contradictory:

We'll call the particular pigeon hole, the one you happen to be examining, the present. The earlier pigeon holes, the ones for which you find substantially correct statements, are what we will call the past. The later pigeon holes, the ones for which there isn't too much in the way of correct statements, we call the future . . . the actual world is very much like this. Instead of pigeon holes we talk about states.

With that word 'states', Hoyle introduces the ideas of quantum theory, although he is careful not to alarm his readers by telling them this. He simply tells them that there is a division into a large number of states, and choosing any one of them constitutes the present. He then suggests that a beam of light could dance about at random over the stack of pigeon holes, illuminating first one hole then another in any order. If the light 'switches on' consciousness then that consciousness will always be aware of the past and the future in just the same way we are, and will experience a steady flow of time.

Now imagine that there are two separate sets of pigeon holes, each being illuminated in this way by the same light, switching at random not just among the pigeon holes in one stack, but between one stack and the other. For the purposes of the story, Hoyle suggests that each set of pigeon holes corresponds to a different human awareness: 'one set of pigeon holes is what you call *you*, the other is what I call *me*.' But in the context of the world view I am describing in this book, one stack of pigeon holes would correspond to one universe, and the other to another universe. There would be an infinite number of stacks in the Multiverse, and we would not have to worry about dancing spots of light switching anything on. Each quantum state just *is*, complete with its more or less accurate memories of the past and its vague extrapolations into the future.

The idea at the heart of Hoyle's book is that all different times are equally valid quantum states. Although he never developed this idea in his scientific work, it has been developed in detail by other scientists, notably including David Deutsch and another Oxford physicist, Julian Barbour. It was Deutsch who summed up these ideas succinctly in the memorable statement 'other times are just special cases of other universes', although perhaps the qualification 'special' is not really needed.

The best way to get a grip on these modern versions of the idea is to start, as Hoyle did, with the block universe model. The name comes from the idea of the universe as laid out in a four-dimensional 'block' like a slab of cake. Nothing ever moves within spacetime, but what we think of as instants of time are slices though the block. If the time dimension is represented by one direction along the side of the block, the viewpoint of what we think of as a stationary observer is represented by slices at right angles to the time axis. What we think of as moving observers see things differently because their instants of time correspond to slices at different angles through the spacetime, the exact angle depending on their speed. All of this matches the description provided by Einstein's equations of how motion affects an observer's view of the universe, with moving clocks running slow and moving objects shrinking in the direction of their motion, with the important proviso that causes always precede the effects they cause – you cannot, for example, slice spacetime in such a way that an observer sees that the light in a room goes on before someone flips the switch that turns the light on.

In its simplest form, the block universe model seems to remove the possibility of free will, since the future is just as fixed as the past – but, as we shall see, quantum physics changes that. Leaving this aside for the moment, each slice through spacetime contains a history of the past, just as in Hoyle's pigeon-hole analogy, and Deutsch points out that if spacetime were sliced very thinly, shuffled like a pack of cards and glued back together with all the moments of time jumbled up, the inhabitants of the spacetime would notice no difference. It's exactly like the jumbled-up pages in the cosmic library. Whichever one you are reading, or experiencing, you know about the pages or slices that are in what we call the past, and you don't know about the pages or slices in the future. Deutsch calls these slices through spacetime snapshots, and says:

Any one of the snapshots, together with the laws of physics, not only determines what all the others are, it determines their order, and it determines its own place in the sequence. In other words, each snapshot has a 'time stamp' encoded in its physical contents.

Barbour has another way of describing the collection of snapshots – there is no need for them to be glued together to make a block of spacetime at all; they can simply be in what he calls 'a heap', a jumbled collection of moments.*

Now add in the concept of the Multiverse. If the Multiverse were just a collection of spacetime blocks, we could imagine them stacked up side by side, or one on top of the other – the old idea of parallel universes. But if each universe is just a jumbled heap of moments, then the Multiverse is also a jumbled heap of moments, a bigger heap made by jumbling all the possible states of all the possible universes together. There is no way to tell from outside (not that there is any outside!) which universe any particular snapshot corresponds to, any more than it is possible to tell which time any particular snapshot corresponds to. That's why Deutsch says that other times are just special cases of other universes (and why I say they are not even particularly special). All possible quantum states exist, corresponding to all possible moments of time in all possible universes.

And the quantum picture gives us back our free will. If there is only one universe in a frozen block of spacetime, we have no free will. But if all possible futures exist, then we have in a real sense a choice of futures. Anything possible can happen, but it is our decisions that

---

* The most succinct account of Barbour's variation on the theme can be found in Lee Smolin's book *The Life of the Cosmos*.

determine which of those futures we experience. When I wrote *In Search of Schrödinger's Cat*, I summed up the nature of quantum physics by quoting from John Lennon: 'Nothing is real.' With what I have learned in the past twenty-five years, I would now change my tune, and agree with Deutsch and Barbour that *everything* is real. But if other times are as real as other places, and since we can travel to other places, might it be possible to travel to other times?

TIMESLIPS

The best description of space and time that we have is the general theory of relativity. It has passed every test that it has been presented with, accurately describing the observed behaviour of distorted spacetime in the presence of matter under conditions ranging from the extreme of two neutron stars orbiting one another and spiralling together as they lose energy by gravitational radiation, to the subtleties of the barely detectable wobbles of a weightless spinning gyroscope in a spacecraft orbiting the Earth. The most remarkable feature of the general theory is that it allows for the possibility of time travel. There is nothing in our present understanding of time which makes it impossible for a traveller to follow a path through spacetime which takes the traveller on a journey that ends back where the journey started, in both space and time. And that means travelling backwards in time for part of the journey.*

Of course, it would be extremely difficult to build a machine capable of taking us on such a journey, and the technology involved is far beyond anything available today. But the point is that it is not impossible. I have discussed the details elsewhere;† here, I want to concentrate on the implications for our understanding of the Multiverse provided by the possible existence of a particular kind of time machine, called a wormhole. This inevitably involves using everyday language corresponding to our human perception of a flow of time

---

* The general theory also tells us that it is impossible to travel back in time to a moment earlier than the moment that the time machine was invented, which may explain why there are no visitors from the future on Earth today.
† *Time Travel for Beginners*, John Gribbin and Mary Gribbin, Hodder, 2008.

from the past to the future; but the paths through spacetime that I describe should really be thought of as just that – paths linking different quantum states within the Multiverse, not actual journeys along those paths.

The heart of the mystery of quantum physics is encapsulated in the experiment with two holes. The time-travel version of this experiment also involves two holes, but these are the opposite ends of a wormhole, which is a tunnel through space and time. According to the known laws of physics, a technological civilization sufficiently advanced to manipulate black holes could set up such a wormhole in such a way that the two 'mouths' of the tunnel were side by side in space, but one mouth was in the past relative to the other mouth. Alternatively, it is possible that such wormholes might exist naturally in the universe, waiting to be discovered; unfortunately, the spacefaring technology needed to find a natural wormhole is almost as advanced as the technology required to make one. But if you had such a wormhole – natural or artificial – it would be possible to send a test object into one hole in such a way that it emerged from the other hole *before* it entered the first hole.

Since this is only an imaginary experiment, we can look at the implications by imagining such a setup on a conveniently small scale, with a wormhole just big enough to send a ball-bearing through, being studied in a laboratory somewhere. This simple example is sufficient to highlight the philosophical problem with time travel – the possibility of so-called paradoxes. What happens if the ball-bearing leaving the past mouth does so in such a way that it collides with its earlier self so that the earlier version of the ball is deflected, and never enters the wormhole? Then it cannot have emerged in time to make the deflection, so it must have gone into the hole, so it must have been deflected ... and so on. But there are also possible trajectories in which the ball emerging from the past mouth strikes its younger self in just the right way to knock it in to the future mouth and set up a self-consistent loop. After the interaction, the version of the ball that has emerged from the past mouth then goes on its way, obeying the standard laws of physics. Researchers such as the American Kip Thorne and the Russian Igor Novikov have proved mathematically that for every paradoxical possibility of this kind there is always an infinite number of self-consistent 'solutions' to the puzzle.

They went on to explain what would be seen if a real ball encountered a real wormhole set up like this, from the perspective of the Many Worlds Interpretation of quantum physics. From a distance, it is impossible to tell exactly which path through the wormhole the ball is following. All you see is the ball approaching the two holes, interacting with them in some way, then leaving the vicinity of the two holes. This turns out to be exactly the equivalent of the behaviour of a photon or an electron going through the double-slit experiment. It approaches the two holes, interacts with them in some way, and then emerges on the other side. Thorne and his colleagues have found that the only workable explanation of what happens when a ball is confronted by a situation like this is that all the possible paths interfere with each other to produce what looks like a single final state. Novikov and his colleagues have proved that only self-consistent solutions to the equations satisfy the rules of quantum physics. Paradoxical paths get cancelled out by interference. In Deutsch's language, all the universes are the same up to the point where the time travel interaction occurs, then the stories told in the different universes diverge from one another, but on the other side of the interaction the stories are once again the same, and always tell a self-consistent story. Or perhaps the universes are not all identical on the other side of the time experiment; maybe it really does mark a place in the Multiverse where the histories diverge, but the stories must still all be self-consistent.

The Many Worlds Interpretation also explains how we can avoid paradoxes when we enlarge our time-travel perspective from the scale of ball-bearings to the human scale. The supposed paradoxes are familiar from many science fiction stories. Suppose I build a time machine, go back into the past, and prevent my parents ever meeting one another. In which case I was never born, so my parents did get together, and so on. Leaving aside the fact that a time machine cannot travel back to a time before it was built (perhaps I found a time machine left on Earth thousands of years ago by aliens), the puzzle is simply resolved in the Multiverse. I do indeed travel back in time and I do indeed stop my parents meeting – but this happens in a different universe. In the world where I came from, my parents do meet, and I was born. Science fiction writers sometimes use this kind of idea within the framework of the image of the universe as a many-branched tree, suggesting that travel backwards in time somehow involves 'sliding down the trunk' of the tree, and that any changes caused by the presence of the time traveller in the past result in a 'new' branch of history growing out from the main trunk. But as I have stressed, there is no main trunk, and no 'new' branches, just different universes with different overall histories.

Suppose I were to travel back to 1066, and participate covertly in the Battle of Hastings on the side of the Anglo-Saxons to ensure that the Normans lost. This would not *create* a new history in which England remained truly English. Such a history already exists – there must be many universes in which William of Normandy lost the Battle of Hastings. Indeed, the odds against William were so enormous that it is almost tempting to think that in our universe *he* must have had help from a friendly time traveller. If there is a time machine, or wormhole, old enough to allow this particular journey, then there is indeed a version of history in which a traveller in a universe very like our own universe (but probably not me!) does travel 'back in time' to the quantum state of the Multiverse corresponding to one of those histories in which the Anglo-Saxons are victorious. On the other hand, a traveller from the future visiting us today would not be from *our* future, although he or she (or it) might be from a universe with a very similar future to ours. The traveller could warn us of impending disaster in time for us to divert it, even though in the traveller's own 'home' universe the disaster has already happened.* But nothing is really changed by any of this; the landscape of the Multiverse remains, with different paths linking different states. Once again, Deutsch sums things up neatly:

Past-directed time travel would inevitably be a process set in several interacting and interconnected universes. In that process the participants would in general travel from one universe to another whenever they travelled in time.

* For an intriguing variation on this theme, see Gregory Benford's *Timescape* (Simon & Schuster, 1980).

BROADER HORIZONS

There is a sense, though, in which even Deutsch's horizons are too limited. In his vision of the Multiverse, there are no universes in which the values of the 'coincidental' numbers discussed in Chapter Two are different from the values they have in our Universe. This is reasonable enough when discussing ideas such as quantum computers and time travel. The kind of interference between worlds required to make such things possible will only work between copies of the universe that are sufficiently similar to one another, and that would certainly include having the same value for numbers like the values of the energy levels in carbon nuclei, or the strength of gravity. But from a broader perspective – in a sense, if we look farther away across the Multiverse, although as the metaphor of the library makes clear that is not an accurate way of looking at things as far as the quantum physics is concerned – we can imagine an infinite variety of universes with different values of the physical constants. Within each subset of this infinite Multiverse (each set of universes where the constants are the same), the whole scenario described by Deutsch, Barbour and others is a good description of reality. And each of those subsets can itself be infinite, because infinity is very big indeed.

This leads to an obvious question. Restricting ourselves just to the infinite subset of universes where a universe very like the one we see around us emerged from a Big Bang some 14 billion years ago, *how* could the numbers discussed in Chapter Two have turned out differently? Reflecting the fact that nobody knows exactly what happened to 'set' those numbers in our Universe, there are several possible explanations on offer. Any one of them may be right – or, reflecting the nature of the Multiverse, they may all be right. They could all be wrong; but they are the best explanations of the cosmic coincidences around at present, and well worth taking a look at. I'll start with the simplest, oldest, but perhaps least satisfying, possibility – another take on infinity.

4

Infinite in All Directions

The idea that the Universe is infinite goes back more than four hundred years. Thomas Digges was, as I have mentioned, the first person to suggest in a scientific publication that the Universe might be infinite. Digges' father, Leonard, invented the theodolite some time in the early 1550s, and used it both in his work as a surveyor and as a telescope; his invention was kept secret for a long time because of its value both in civil work and in military surveying. Thomas Digges followed in his father's scientific footsteps, and his ideas about infinity actually appeared in 1576 as an appendix to a revised and expanded version of a book originally written by his father, who had died in 1559. Digges asserted that the Universe is infinite in its spatial extent. In his own words, '[the] orb of stars fixed infinitely up extendeth itself in altitude spherically' around the Sun.

Although the idea was not taken up at the time, it marked a profound change from the way the Ancients thought about the Universe. Greek philosophers, for example, were uncomfortable with the idea of a spatial infinity, for understandable reasons. Infinity is not just a very big number, like the biggest number you can imagine plus a little bit more; it is something else entirely. The simplest way to begin to appreciate the weirdness of infinity is first to consider all of the integer numbers – 1, 2, 3, 4, 5, and so on. They form a mathematical set, which goes on forever, so it is an infinite set. Whatever number you can think of, there is always a bigger number you can get by adding 1. Now think about the set of all the even numbers – 2, 4, 6, 8, and so on. At first sight it seems that this set is smaller than the first set, because you have left out all the odd numbers. Surely it contains only half as many numbers as the first set? But you can make each number in the second set simply by doubling each of the integers in the first set. So every number in the first set pairs up with a number in the second set. 1 pairs up with 2, 2 pairs up with 4, 3 pairs up with 6, and so on. In that case, it looks as if both sets are the same size, because every number in one set has a unique counterpart in the other set. You have two infinities, one containing the other, but both the same size as each other!

In spatial terms, there is room in an infinite universe not only for everything possible to happen, but for an infinite number of infinite universes, in each of which anything possible can happen an infinite number of times. No wonder the Ancients recoiled from the idea, and preferred to imagine that the Universe is finite and bounded by a celestial sphere. Aristotle, for example, was happy to consider the idea of infinity in mathematics, but said firmly, in his *Physics*, that 'there will not be an actual infinite.' On the other hand, philosophers such as Aristotle were unhappy with the idea that the Universe might have had a beginning, because that would mean a beginning, and perhaps an end, to time. So they were willing to accept the idea of an infinite duration of the Universe in time. This is almost exactly the opposite of the standard modern cosmological view of the visible Universe. Cosmologists today are quite happy to consider the idea that the Universe is infinite in space, but their standard models start from the Big Bang at a definite moment in time, 13.7 billion years ago. But if our Universe is just one component of the Multiverse, the Multiverse itself may be infinite in all directions – in time as well as in space.

As the astronomer James Jeans wrote in his 1930 book *The Mysterious Universe*, 'if the universe goes on for long enough, every conceivable accident is likely to happen in time.' This statement of what is to some extent the obvious can be put on a secure scientific footing by taking a look at what we understand about time, and how our perception of the flow of time relates to the ordered nature of the expanding Universe we see around us.

ARROWS OF TIME

The relationship between order and time is clear from our everyday experiences. The classic example is an ice cube melting in a glass of water. Before the ice melts, the contents of the glass are in an ordered state, with ice and liquid water distinct from one another. You would need a certain amount of information to describe the situation, either in words or mathematically. After the ice melts, there is less order – which means less information, and less complexity. All you have is a simple liquid. It takes fewer words, or simpler equations, to describe the situation. The description 'a glass of water' replaces the description 'a glass of water with an ice cube in it'.

What we mean by 'order' in this sense is related to the more precise scientific concept of entropy, in such a way that decreasing order corresponds to increasing entropy. Left to their own devices, with no influence from the outside world, things like ice cubes in glasses of water always change in the sense that the amount of entropy increases, while the amount of complexity decreases. The caveat 'left to their own devices' is crucially important, however, because this rule applies strictly only to what are called closed systems. The complexity of life on Earth is an obvious example of a system where entropy has decreased as time has passed, but this is only possible because life on Earth feeds off an energy supply from outside – from the Sun. The entropy of the Solar System as a whole has indeed been increasing during the billions of years that it has taken for the present complexity of life on Earth to evolve.

Time comes into the story because we all have a clear idea of the way time seems to flow. You start with a glass of water containing an ice cube, and end up with a glass containing just water. You never see an ice cube spontaneously forming itself out of the molecules in a glass of water, and another way of expressing the law of increasing entropy (known as the second law of thermodynamics) is that heat always flows from a hotter object to a cooler object, never the other way. There is a clear distinction between what we call the past and what we call the future. But remember that although we *perceive* a flow of time that does not necessarily mean that there *is* a flow of time. There is without doubt an arrow of time* that *points* from the state of lower entropy towards the state of higher entropy; but this does not mean that there is an arrow of time which *moves* in the direction of increasing entropy. It's like the difference between the arrow on a magnetic compass needle, dutifully pointing North but sitting still in one place, and the arrow, fired from a bow, that actually travels North, or in some other direction. There is nothing in the idea of entropy as an indicator of the arrow of time that conflicts with the ideas about the nature of time described in Chapter Three.

We also have another arrow of time, at least in our part of the Multiverse, provided by the Universe itself. The past is in the direction of the Big Bang; the future is the direction in which we see the Universe getting bigger. This is crucially important to an understanding of why we perceive a thermodynamic arrow of time in our Universe; but the simple thermodynamics is relevant to the possible existence of universes like our Universe within the infinite Multiverse. This is because of an apparent conflict between the simple laws that govern the behaviour of things like atoms and molecules and the second law of thermodynamics – a puzzle that was more puzzling before astronomers discovered that the Universe is expanding, but is still relevant today. The puzzle is that at the level of atoms, molecules and subatomic particles there is no arrow of time.

The usual way to make this clear is to think about the molecules of air bouncing around in a room, or in a sealed box. To make things even simpler, we can think about a box containing atoms of the inert gas neon. This doesn't react chemically, removing the possible complications of chemical interactions involving the atoms. The atoms bouncing around inside the box, colliding with each other and bouncing off the walls of the box, behave almost exactly like hard little spheres, tiny pool balls ricocheting around inside the container. Unlike the behaviour of real pool balls travelling across a table, these collisions do not involve friction, and as long as the box is kept at a uniform temperature and pressure, there is no way to determine an arrow of time by monitoring the collisions. Two atoms move together,

---

* The expression was introduced by the physicist Arthur Eddington in 1928, in his book *The Nature of the Physical World*.

collide and bounce off one another, and everything about the collision obeys all the laws of physics; if you then reversed time and watched the collision running backwards, everything about it would still obey the laws of physics. And, unlike the case of an ice cube melting in a glass of water, if you took a series of snapshots of the state of the gas inside the box and jumbled them up, there would be no way to tell which order the snapshots had been taken in. There is no trace of an arrow of time in the behaviour of the atoms of gas in the box.

In order to see an arrow of time, we have to set the system up in a state which is not in equilibrium, equivalent to dropping an ice cube in a glass of water, and then watch how it settles down into a steady state.

Imagine that we have a box which is divided in two by a sliding partition. On one side of the partition, half the box contains neon gas; on the other side of the partition, the box has been pumped free of gas and is as empty as we can make it. We all know what will happen when the partition is slid out of the way – gas will spread out to fill the whole box. Now, if we look at the same thing in reverse it seems silly; in real life we do not see a box full of gas in which all the atoms suddenly move to one end of the box, leaving a vacuum behind them. We seem to have found an arrow of time operating even at the level of atoms and molecules. But things aren't quite that simple. Even while the gas is expanding to fill the whole box, every collision between atoms obeys the reversible laws of physics, and every collision would look perfectly natural if we could view it running in reverse. So where, between the level of a simple collision between two atoms and the expansion of a gas made of very large numbers of atoms to fill a box, does the arrow of time emerge?

The simple answer is that it doesn't. There is, indeed, nothing in the laws of physics which says that all the gas in the box can't just happen to move into one half of the box, leaving a vacuum behind it, before spreading out again to fill the whole box; it's just that such an occurrence is extremely unlikely. There are an awful lot of atoms in a box of gas, and any one of them could be anywhere in the box at any instant, so on average they will be spread out more or less evenly through the box. In more technical language, there are very many states corresponding to an even spread of atoms, even though any particular atom may be anywhere in the box. There are far fewer states corresponding to all the atoms being in one end of the box at the same time, so it isn't likely that we will see this. But in 1890 the French physicist Henri Poincaré showed that an 'ideal' gas, trapped in a box where all the collisions involve no loss of energy, must eventually pass through every possible state that is consistent with the law of conservation of energy. Every possible arrangement of atoms inside the box must happen some time, and if we wait long enough we will see all the atoms moving to one end of the box. In other words, the atoms must eventually return to their starting points, and the original situation will recur – provided we wait long enough.

There's the rub. The time it takes, statistically speaking, for the atoms to return to their original state is known as the Poincaré cycle time (or Poincaré recurrence time), and it depends on the number of atoms involved. Roughly speaking, the Poincaré cycle time is 10^N seconds, where N is the number of atoms involved. For a box containing just 10 atoms, this would be 10^{10} seconds, or a bit more than 300 years. A small box of gas might actually contain 10^{23} atoms, so the recurrence time is 10 *to the power of* 10^{23} seconds. But the age of the Universe (the time since the Big Bang) is only about 10^{17} seconds. The difference between 17 and 10^{23} (a 1 followed by 23 zeroes) is a measure of the statistical (im)probability of seeing the gas move towards one end of the box during the entire lifetime of the Universe so far; this utterly insignificant probability, associated with a relatively small Poincaré cycle time, provides the standard answer to the puzzle of how a world that is timeless on the small scale can have an apparent arrow of time on the large scale – that arrow of time is merely a statistical illusion. If you could watch a glass of water for long enough – many, many times the age of the Universe – you would indeed eventually see the bulk of the water getting warmer while a lump of ice formed in the middle.*

This is a persuasive explanation of the perceived arrow of time if the Universe does indeed have a finite age which is much less than a typical Poincaré cycle time. But it is less persuasive if the Multiverse is infinitely old. Although he did not use the term Multiverse, the

---

* Or it *could* happen next time you look at a glass of water, with odds of more than 10^{23} to 1 against.

Austrian physicist Ludwig Boltzmann pointed out the fatal flaw in Poincaré's argument in the mid-1890s. In essence, his suggestion was that our visible Universe is a temporary low-entropy bubble that has emerged just by chance in an infinite, eternal and timeless high-entropy world – the cosmic equivalent of all the gas moving up into one end of the box. If so, the Universe is indeed just one of those accidents referred to by Jeans. The argument fell from favour with the discovery of the expansion of the Universe and evidence for its origin in the Big Bang a finite time ago; but it is once again worthy of serious consideration in the context of the Multiverse. Its full force can be seen by considering what the 'natural' state of the Universe ought to be, according to the laws of thermodynamics – something that worried scientists as long ago as the 1850s.

THE HEAT DEATH OF THE UNIVERSE

The strangest feature of the Universe is that it contains bright stars scattered across a dark sky. All those stars are busy pouring out heat energy into the cold Universe, in line with the rule that heat flows from a hotter place to a cooler place. If this had been going on forever, the space between the stars should have become full of radiation with the same temperature as the stars themselves, and the Universe would be in equilibrium. Alternatively, if we wait long enough the stars will burn out, and since they cannot produce enough energy to heat the entire Universe to the temperature of a star, they will end up as cold cinders in equilibrium with their cold surroundings. Either way, if there is no contrast between hot and cold places no heat can flow, the Universe is in thermodynamic equilibrium (or thermal equilibrium, which is much the same thing) and nothing interesting could happen.

What we now know as the second law of thermodynamics was introduced into scientific discussion by Rudolf Clausius, a German physicist, in a paper published in 1850. Almost casually he wrote that 'heat always shows a tendency to equalize temperature differences and therefore to pass from *hotter* to *colder* bodies.'* Over the next

* His italics.

few years, Clausius and other researchers refined the idea and put it on a secure mathematical footing; but the 1850 paper is regarded as marking the beginning of the science of thermodynamics.

The first person to relate these ideas to the puzzling existence of hot stars in a cold Universe was William Thomson, who was born in Belfast but spent most of his life in Scotland, and who was later elevated to the peerage as Lord Kelvin, the name by which he is better known today. Among his many achievements, Thomson independently arrived at some of the same conclusions as Clausius; the 'absolute' scale of temperature, based on thermodynamic principles, is named the Kelvin scale in his honour. In 1852, Thomson published a scientific paper with the title 'On a Universal Tendency in Nature to the Dissipation of Mechanical Energy', which offered an early version of the idea that the Universe is running down. This aroused considerable interest and discussion among the experts, and in 1862 Thomson published a paper, 'On the Age of the Sun's Heat', in which he pointed out that 'if the universe were finite and left to obey existing laws', then 'the result would inevitably be a state of universal rest and death.' The term 'heat death' (or 'thermal death') was introduced to describe this ultimate fate of the Universe two years later, by the German Hermann Helmholtz; both Thomson and Helmholtz realized that the fact that the Universe is running down in this way implies that it used to be in a state of lower entropy, which, they thought, must have been produced by some cause outside the laws of physics known to them. By the 1890s, the idea of the heat death of the Universe had become such a pervasive concept that it even features in H. G. Wells' classic story *The Time Machine*. But by then, Poincaré and Boltzmann had arrived on the scene.

EVERY CONCEIVABLE ACCIDENT

Boltzmann's key contribution to this debate came in a paper published in *Nature* on 28 February 1895, under the rather bland title 'On Certain Questions of the Theory of Gases'. There, he repeated a point that he had emphasized before, that the 'so-called Second Law of Thermodynamics' is actually only a statement of probability. He then discusses the probability that a box containing a mixture of nitrogen and oxygen gases will unmix itself, with all of the oxygen going to one end of the box and all of the nitrogen going to the other end, in terms of a measure related to probability that he refers to by the letter H; this is roughly the opposite of entropy, so 'the probability that H decreases is always greater than that it increases.' Boltzmann uses the example of a die that is thrown 6,000 times: 'we cannot prove that we shall throw any particular number exactly 1,000 times; but we can prove that the ratio of the number of throws in which that number turns up to the whole number of throws, approaches the more to 1/6 the oftener we throw.' But improbable deviations equivalent to getting a run of the same number coming up on several successive throws of the die can happen. These are equivalent to peaks in the value of Boltzmann's H parameter; Boltzmann uses statistical arguments to show that it is very unlikely that the mixed box of gas will unmix itself, and confirms the idea of the heat death of the universe, which 'must tend to a state where . . . all energy is dissipated'. Then, just when his argument seems to be complete, comes a sting in the tail:

I will conclude this paper with an idea of my old assistant, Dr Schuetz.

We assume that the whole universe is, and rests for ever, in thermal equilibrium. The probability that *one* (only one) part of the universe is in a certain state, is the smaller the further this state is from thermal equilibrium; but this probability is greater, the greater is the universe itself. If we assume the universe great enough, we can make the probability of one relatively small part being in any given state (however far from the state of thermal equilibrium), as great as we please. We can also make the probability great that, though the whole universe is *in thermal equilibrium*, our world is in its present state. It may be said that the world is so far from thermal equilibrium that we cannot imagine the improbability of such a state. But can we imagine, on the other side, how small a part of the whole universe this world is? Assuming the universe great enough, the probability that such a small part of it as our world should be in its present state, is no longer small.

If this assumption were correct, our world would return more and more to thermal equilibrium; but because the whole universe is so great, it might be probable that at some future time some other world might deviate as far from thermal equilibrium as our world does at present. Then the afore-mentioned H-curve would form a representation of what takes place in the universe. The summits of the curve would represent the worlds where visible motion and life exist.

So Boltzmann fluctuations should really be called 'Schuetz fluctu-ations'! Either way, this is an astonishing passage to find in a paper published in 1895, which is directly relevant to modern ideas about the Multiverse, if we replace Boltzmann's term 'universe' with our 'Multiverse,' and his 'world' with our 'Universe'. It even includes implicit, if unconscious, anthropic reasoning – observers like us can only exist in fluctuations like these, so it is no surprise that we find ourselves living in such a fluctuation.

But Boltzmann soon made the idea his own, and defended it vigorously, even if the seed had been planted by Schuetz. In 1897, he wrote:

This viewpoint seems to me to be the only way in which one can understand the validity of the Second Law and the heat death of each individual world without invoking a unidirectional change of the entire universe from a definite initial state to a final state. The objection that it is uneconomical and hence senseless to imagine such a large part of the universe as being dead in order to explain why a small part is living – this objection I consider invalid. I remember only too well a person who absolutely refused to believe that the sun could be 20 million miles from Earth,* on the grounds that it is inconceiv-able that there could be so much space filled only with aether and so little with life.

There is, though one point that Boltzmann overlooks, where he refers to other 'worlds' like ours forming 'at some future time'. In the wider universe he envisages (a better term might be the meta-universe), there is no time. In thermodynamic equilibrium, there is no way to distinguish the past from the future, any more than a series of snap-shots of a box of gas in thermal equilibrium can be jumbled up and

---

* In fact, the Sun is 93 million miles (some 150 million km) from Earth. This translation is from the book *Kinetic Theory*, by S. G. Brush, Pergamon, Oxford, 1966. The 'aether' was a hypothetical fluid filling what we regard as empty space; the need for such a fluid was removed by the discovery of the special theory of relativity.

then arranged in the order they were taken simply by looking at the distribution of atoms on each snapshot. If we are living in a fluctuation within such a meta-universe, all that can be said about the meta-universe is that it exists, and that within it other fluctuations exist. The arrow of time (or arrows of time) only exist within those fluctuations.

There's another puzzle, which Boltzmann addresses but which worries a lot of people today. Why do we live in such a large fluctuation from thermodynamic equilibrium? Boltzmann was happy that no matter how big our 'world' (Universe) is, 'assuming the universe great enough, the probability that such a small part of it as our world should be in its present state, is no longer small.' There may be smaller fluctuations as well, but so what? The puzzle, as it is usually expressed today, is that from the perspective of the meta-universe it should be much easier to make a much smaller fluctuation, starting out from thermodynamic equilibrium.

To take an extreme example, if a fluctuation can occur that is as large and complex as the entire Universe, it ought to be much easier, and therefore much more likely, that a fluctuation could produce the room you are sitting in, all it contains, yourself, complete with all your memories, and this book. It could have happened a second ago, and it could all disappear before you finish reading this sentence. I'm sure you will be glad to learn (if you have indeed survived to read this far) that there is a major flaw in this argument, and it turns out, as I shall explain later, that it is much easier to make a big bang out of a Boltzmann fluctuation than it is to make, say, a human brain in one step. But even without this reassuring insight, it is still true that in an infinite universe anything is possible, and no matter how unlikely the Universe we live in is, it is bound to happen somewhere. The odds against a fluctuation the size of our Universe existing might be so enormous that we cannot truly comprehend them – but *any* number is, literally, infinitesimally small compared with infinity. Which raises the interesting question of just how likely it is that there are other copies of ourselves somewhere in the meta-universe, and how far away they might be.

TIME AND DISTANCE

To put this in perspective, it's worth spelling out exactly how a Boltzmann-type fluctuation makes pockets of order in a disordered meta-universe. The appropriate nineteenth-century image would be of a sea of particles and radiation in equilibrium, with bubbles of order appearing in the chaos. But those bubbles cannot just appear instantaneously, fully formed. They have to develop as fluctuations away from the state of equilibrium. Particles (and radiation) have to come together in just the right way to make, in the case of a fluctuation like the one we live in, stars, planets and people. From the perspective of our everyday perception of the flow of time, this would be exactly like time running backwards, with, for example, radiation converging on stars and burrowing deep into their interiors, where the effect of the radiation would be to break apart nuclei of elements such as carbon and convert them into helium. It is from this perspective that we say that it is much more likely that this process should produce a small, short-lived bubble the size of your room than that it will produce a universe as big as the one we see around us; and it is the same perspective which suggests that it is much more likely that the process has only just stopped and gone into reverse, heading back towards the chaotic, low-entropy state, than that it really did proceed all the way to the Big Bang before reversing and beginning the journey back towards disorder.

As we have seen, it is very difficult to say what time is, and all of this description is essentially the view of some cosmic observer who stands outside of time and watches the meta-universe evolve; from the point of view of anyone living in the Boltzmann fluctuation, however, there is no reason to think that they would experience 'time running backwards' while the fluctuation was forming. Although this intriguing idea has led to several entertaining science fiction stories, notably Philip K. Dick's *Counter-Clock World*, all of the arguments used in the previous chapter would apply with full force in both halves of the fluctuation. While the fluctuation was forming, there would be a series of time slices (equivalent to Hoyle's pigeon holes) with a more ordered universe in some slices and a less ordered universe in others. If our perception of the flow of time is indeed based on arranging the slices in such a way that the order is in the 'past' and the disorder is in the 'future' then the same rules apply in both halves of the fluctuation. We could be living in the bubble during its formation, and never know. In other words, from the lowest entropy state that the fluctuation reaches, no matter how high or how low it may be, the arrow of time points in both directions 'uphill' towards the high-entropy state, as seen by the observer standing outside of time. *Everywhere* in the bubble, the arrow of time points towards the high-entropy state.

With this classic version of the Boltzmann fluctuation idea, we still have to accept that there may be fluctuations the size of your room, or the size of an entire galaxy, or any other size. Modern ideas about the nature of the Big Bang and of the expanding Universe give us a rather different perspective, which I describe in Chapter Five. But even in the modern framework, we still have to accept that in an infinite Multiverse there will be copies of you, me, planet Earth, and everything up to and including the entire visible Universe. Not necessarily another version of you in an identical room but surrounded by eternal chaos; rather, another version of you in an identical room on a planet very similar to the Earth orbiting a star rather like the Sun in a galaxy not unlike our Milky Way, and so on.

Without worrying too much at this stage about how the different universes arise, it is possible to get a rough idea of how far the copies are from each other simply by looking at the number of ways in which the atoms and particles that make up your body, or the Earth, or a galaxy, or the entire Universe can be arranged. If you imagine (or calculate), for example, how many different arrangements of atoms there could be using the atoms in your body, and then think of each of these 'bodies' side by side in a line, with no repetition, how long would the line be before you ran out of permutations and had to start again? Max Tegmark, of the University of Pennsylvania, has done the number-crunching for us, and put it in the perspective of the size of the visible Universe.

The time since the Big Bang is just under 14 billion years, so in that sense the Universe is a little less than 14 billion years old, and the most distant objects we can see are viewed by light which left them a little under 14 billion years ago. But while that light has been travelling through space to us, space itself has been expanding and the Universe has grown, so that the most distant visible objects are not now about 14 billion light years away but a little more than 40 billion light years away – 4×10^{26} metres from us. This is the radius of a bubble of space around us that is called the Hubble volume, in honour of the astronomer who discovered that the Universe is expanding. Other versions of yourself living on other planets elsewhere in the Multiverse are surrounded by their own Hubble volumes, containing everything that they could see if they had perfect telescopes. But there is no reason to think that these Hubble volumes overlap one another. Far from it – literally.

Assuming only that space is infinite and that it is uniformly filled with matter on the large scale,* Tegmark calculates that your nearest 'twin' is living on a planet about 10 *to the power of* 10^{29} metres away. This is an incomprehensibly large number. Tegmark likes to say that it is 'more than astronomical', since it far exceeds the radius of the Hubble volume, which contains everything astronomers can study with their telescopes. But it is still infinitesimal compared with infinity; the size of the number, Tegmark emphasizes, does not make your twin any less real.

Using the same assumptions, there should be a spherical volume of space, 100 light years in radius and with contents identical to those of the region of space 100 light years in radius centred on the Earth, at a distance of 10 to the power of 10^{91} metres away; and there should be an identical copy of our entire Hubble volume at a distance of 10 to the power of 10^{115} metres. And there must be many more slightly imperfect copies of our Universe, variations on the theme not unlike the different versions of quantum reality in the Everett interpretation, where the copies of you make different choices that affect their future lives. As Tegmark puts it, 'everything that could in principle have happened here did in fact happen somewhere else,' such as you winning the lottery, or life on 'Earth' being destroyed a week ago by the impact of a large asteroid.

---

* Evidence from observations of the cosmic background radiation shows that there are no structures in our Universe more than 10^{24} metres across, which is what Tegmark means by the large scale.

Tegmark refers to this kind of situation, in which there are other copies of everything in an infinite, expanding meta-universe where everything emerged from the same Big Bang and everything obeys the same laws of physics, as a 'Type I' Multiverse. As the name implies, there are other types of Multiverse, including the Everettian parallel universes of quantum physics (which Tegmark classifies as 'Type III') and different possibilities that I will describe later. But this is the simplest variation on the Multiverse theme, and Tegmark argues that the idea of a Level I Multiverse is implicitly built in to the assumptions cosmologists make when talking about their interpretation of observational evidence. For example, observations of the cosmic background radiation show a mixture of slightly hotter and slightly colder patches across the sky. The size of these patches is related to the curvature of space, and the observations strongly support the idea that space is not curved like the surface of a sphere. In the language favoured by cosmologists, the spherical model is said to be ruled out with 99.9 per cent confidence. What this means is that according to the standard calculations hot and cold patches the size we see *could* appear by chance in a spherical universe once in a thousand times. But Tegmark says that this interpretation of the numbers only makes sense if there are (at least) 1,000 other spherical universes that could be investigated. What the '99.9 per cent confidence' really means is that 999 spherical universes out of every thousand would show a different size of spots to the patches we see, and it is unlikely that we live in a very rare spherical universe where the spots fool us into thinking space is flat. That's as far as we can go for now with infinite space; but what about infinite time? Where does time come into the cosmological calculations?

TIME AND THERMODYNAMICS

The key difference between the kind of fluctuation from thermodynamic equilibrium that Boltzmann described and the Universe we live in is that the Universe is expanding. In an infinite, eternal meta-universe of steady-state chaos, it would be quite possible, though extremely rare, for a fluctuation to produce a universe exactly like the one we live in but without the expansion; but as we have already seen, this is much less likely than a fluctuation that produces a single room, or a single planet, or even a naked human brain floating in the midst of chaos. The fact that our Universe began in a Big Bang changes our perspective; as I have hinted, and as I describe fully in the next chapter, as far as fluctuations are concerned making a big-bang expanding universe is far easier than making an isolated human brain. But before I get to discussing how the Universe as we know it was born, the nature of our particular Universe – of our particular *kind* of universe – gives us a slightly different local perspective on thermodynamics and the arrow of time. And the fact that the expanding Universe contains lumps of matter which have been pulled together by gravity makes things even more interesting.

Classical thermodynamics was largely developed in the nineteenth century by pioneers such as Clausius, Kelvin, Boltzmann and the American Josiah Willard Gibbs. It essentially deals with systems in equilibrium. It can tell us, for example, that when the divider in the box of gas I described earlier is removed the gas will spread out to fill the entire box, because the box full of gas is in an equilibrium state. But in spite of its successes, classical thermodynamics is not so good at describing what happens while the gas is spreading out to fill the box – while it is not in equilibrium. In effect, as I have discussed elsewhere,* classical thermodynamics pretends that time does not exist. Things like gas spreading out to fill a box, and far more complicated systems, are described in terms of a succession of slightly different states that are each 'static' but differ from one another by tiny steps. Infinitesimally tiny steps, in fact, which would mean that it would take infinitely long for the gas to fill the box – so there is obviously something wrong with this approach. Although the term 'dynamics' in its name implies change, classical thermodynamics does not really describe change at all, and key concepts such as entropy were worked out from calculations of systems in equilibrium, by comparing one system in equilibrium with another system in equilibrium, or with the same system in a different state, like the box of gas before and after the divider is removed. This doesn't mean that those

* *Deep Simplicity.*

key concepts are wrong; but it does mean that they do not tell the whole story.

In the twentieth century, and especially in the second half of the twentieth century, by building from the work of pioneers such as the Norwegian Lars Onsagar and the Russian-born Ilya Prigogine, scientists developed an understanding of non-equilibrium thermo-dynamics, describing how things change and how complex systems (up to and including the complexity of human beings or the web of life on Earth) can be produced out of simple components, seemingly in defiance of the second law of thermodynamics, by feeding off a flow of energy. All this has made the study of what mathematicians call chaos (not the same thing as the kind of thermodynamic chaos envisaged by Boltzmann as the natural state of the world) and com-plexity one of the most important areas of research today; but it is not necessary to go into all of the details here. The one crucial fact we have to take on board is that our ideas about equilibrium have to be modified in an expanding universe.

Let's get back to the simple example of gas expanding to fill a box. Suppose that instead of having a finite box neatly divided into two halves with a divider that is completely removed we have a very long (perhaps infinitely long), tube with a piston in it, and the piston is steadily pulled along the tube, with gas expanding behind the piston to fill the ever-increasing volume left behind by the piston. As long as the piston keeps on moving, the gas can never settle down into thermodynamic equilibrium. Or imagine that the gas starts out in one corner of an infinitely large box, and expands outwards into the surrounding void. It will keep on expanding forever, always moving towards a state of thermodynamic equilibrium but never quite reach-ing it. Or suppose that we start with a small box jammed full of gas, but the box itself expands so the gas expands with it. Or that we open the lid of the box and let the gas out into an infinite vacuum. These are all reasonable analogies for the way the Universe has expanded away from the Big Bang, and the best evidence we have, involving the dark energy described in Chapter Two, is that the Universe will indeed expand forever, thinning out its material content. Don't worry for now about how it got started – that will become clear in Chapter Five.

This expansion and thinning of the Universe provides us with another arrow of time, quite separate from the arrow defined by the tendency of thermodynamic systems to settle into equilibrium and for entropy to increase. There is clearly a cosmological difference between what we call the past and what we call the future; the past is when clusters of galaxies are closer together than they are today, and the future is when clusters of galaxies are farther apart than they are today. In other words, according to our subjective experience of the flow of time, the cosmological and thermodynamic arrows point in the same direction; when the Universe is bigger more stars will have burned out and everything will be closer to thermodynamic equilibrium, in a higher entropy state.

But you should have noticed something odd about all of this description. We can say that the future is when clusters of galaxies are farther apart – but what are large objects like galaxies doing in a Universe that is moving towards thermodynamic equilibrium and started out as a very smooth gas? We can say that the future is when stars burn out – but how did hot objects like stars, clearly not in thermodynamic equilibrium with their surroundings, get to form in the Universe? The answer to both questions is gravity. Gravity allows irregularities to grow in defiance of the laws of thermodynamics that apply in the absence of gravity. Classical thermodynamics does not include gravity at all, since gravity is not very important for an understanding of the behaviour of things like boxes of gas or ice cubes melting in glasses of water here on Earth. But gravity is overwhelmingly important on the scale of stars, galaxies and the Universe. At least for a time, gravity can turn the rules of classical thermodynamics on their head, increasing the temperature differences between one place and another, and increasing the amount of information needed to describe the Universe, because it has a very curious property. The gravitational energy associated with a lump of matter such as a star, or a galaxy – or, indeed, your body – is *negative*. And this means that a gravitational field contains negative entropy – that gravity provides a sink for entropy.

THE COSMIC ARROW AND THE GRAVITATIONAL SINK

I have discussed the negativity of gravity in other books; but it is such an important feature of the Universe, and so counter-intuitive, that I make no apologies for expanding on it again here. The fact that the energy of a gravitational field is negative is the single most important reason why the Universe we see around us exists, and it simply can't be glossed over.

The way to get a handle on this is to think about how the force of gravity works. It obeys an inverse-square law, which means that the force of gravity operating between any two objects (any two lumps of matter) is proportional to 1 divided by the square of the distance between them. When two (or more) objects are pulled together by gravity, this force makes them move faster, converting gravitational energy into energy of motion – kinetic energy. The kinetic energy of atoms and molecules jostling one another is what we call heat. A star, for example, starts life as a cool cloud of gas and dust, but gets hotter as it shrinks and gravitational energy is converted into kinetic energy. Once the proto-star is hot enough inside, nuclear fusion reactions can begin and it can stay hot until the nuclear fuel is exhausted. But it all begins with heat extracted from the gravitational field as the material collapses into a more compact state. This is what allows hot stars to form in a cold Universe, seemingly in defiance of the second law of thermodynamics. Concentrations of matter are like sinks for entropy.

Now imagine disassembling a star like the Sun into its component atoms or particles, and spreading them out into an infinitely large cloud of material, so that each particle is infinitely far away from its nearest neighbour. Since 1 divided by infinity (let alone by infinity squared) must be zero, there will be no gravitational force operating between the particles. In other words, the energy of their gravitational field will be zero. Obviously we cannot really disassemble stars in this way, but this simple thought experiment gives us some insight into a fundamental fact that emerges from a proper calculation using the equations of the general theory of gravity, which says that, indeed, the zero point of the gravitational energy associated with any amount of matter does occur for all the components separated as widely as possible. This is not an arbitrary choice, like choosing to measure temperatures from the zero points of the Celsius or Fahrenheit scales, but a fundamental truth, like measuring temperatures from the zero point of the Kelvin scale.

Any concentration of matter more compact than an infinitely dispersed cloud (even a cloud of gas containing one hydrogen molecule in every litre of space) must have less gravitational energy than an infinitely dispersed cloud, because, when material falls together energy is removed from the field. We start with zero energy and take some away, so we are left with negative energy. The negative energy of the gravitational field is what allows negative entropy, equivalent to information, to grow, making the Universe a more complicated and interesting place, with hot stars pouring out energy, on which planets like Earth can feed, as they attempt to redress the balance. Eventually, entropy will win – but not just yet.

How much negative energy are we left with? A relatively simple calculation using the general theory of relativity tells us that if any collection of matter (a star, a planet, a person, a universe . . .) with a mass m could collapse all the way down to a mathematical point, a singularity, the energy of the gravitational field associated with that mass would be *minus mc^2*. It would be exactly equal and opposite to the rest mass energy of the matter itself, given by Einstein's famous equation. The two energies would precisely cancel out, meaning that any concentration of matter at a point has zero energy overall.

The first person to appreciate the significance of this was the German Pascual Jordan, who realized that it meant that a star might be made out of nothing, if it appeared at a point, since its negative gravitational energy would be exactly equivalent to its positive rest mass energy. In the 1940s, the physicist George Gamow was based in Washington, and mentioned this astonishing idea to Einstein on a visit to Princeton, where Einstein was working. 'Einstein stopped in his tracks,' says Gamow, 'and, since we were crossing a street, several cars had to stop to avoid running us down.'

Astonishing though the idea is – astonishing enough to bring

Einstein to a halt – there are at least two problems with making a star in this way. The first is that the gravitational force holding the matter together would be so strong that it could never expand to make a star;* the second is that if anything like this did happen it would be concealed from view behind the horizon of a black hole. And then there is the problem that no theory can actually describe what happens at a singularity, only what happens very near to a singularity. But – as even Einstein didn't appreciate in the early 1940s, when the expansion of the Universe had only recently been discovered and the term Big Bang hadn't even been coined in a cosmological context – if you could make a star out of nothing at all you could make a universe out of nothing at all, provided it was very, very small to begin with and you had some mechanism to set it expanding rapidly before gravity could snuff it out again. That is exactly what physicists discovered in the 1980s, and is the subject of the next chapter; from that perspective, we are *inside* the black hole, resolving the other problem as well. But gravity still has more to tell us about the arrow of time.

Even a universe which starts out very nearly perfectly smooth and uniform from a big bang, as ours seems to have done, will become more irregular as time passes and gravity pulls concentrations of matter together. If this process goes on for long enough, eventually a great deal of this matter will end in black holes, the most extreme concentrations of matter possible. Black holes make deep dents in the fabric of space, but even less extreme concentrations of matter distort space in their vicinity. From our human perspective, space itself gets more crumpled up as time passes. This would happen, starting out from a nearly uniform universe, even if space was not expanding, so it is independent of the cosmological arrow of time. The future is when space is more crumpled; the past is when space is smoother. And this crumpling of space gives pause for thought when we try to imagine what the fate of our Universe would be if it did not keep on expanding forever.

---

* At a singularity, a mathematical point, the force would be infinite, which is one reason why physicists don't trust theories with singularities in them.

BOUNCING BACK?

The general theory of relativity provides an accurate mathematical description of a universe that expands away from a big bang in exactly the way that we see the Universe we live in expanding away from 'the' Big Bang. But this description of expanding spacetime is not the only cosmological solution to Einstein's equations. The equations actually describe a variety of universes. Some of them expand faster, some more slowly; some are large, some small; some expand forever; some are destined to fall back upon themselves in a 'big crunch' reminiscent of the Big Bang in reverse. This variety of possible universes may also be relevant to the search for the Multiverse – perhaps all the kinds of universes allowed by Einstein's equations exist somewhere in the Multiverse – but for now I want to focus on the implications for our Universe.

For about 80 years after Einstein came up with the general theory, and for long after the idea of the Big Bang became respectable, nobody knew for sure whether our Universe will expand forever, or whether it must one day recollapse. Either possibility is allowed by the equations, and the fate of the Universe depends on the balance between how much matter it contains, with gravity acting to halt the expansion and put it into reverse, and the possible existence of a cosmological constant, acting like antigravity and encouraging the expansion. The uncertainty in our knowledge about both of these contributions persisted up until the end of the twentieth century, allowing plenty of scope for theorists to speculate about the ultimate fate of the Universe. One of those speculations, dating back to the work of the Russian theorist Alexander Friedmann in the early 1920s, is that what we now call the Big Bang might actually have been a 'big bounce' resulting from the contraction of a previous universe (or the same Universe) following an earlier big bang and expansion to a finite size.

This possibility, allowed by the laws of physics and specifically by the equations of the general theory, would presumably have made Aristotle happy, since it seems to allow for the possibility of a spatially finite universe with an infinitely long history, removing the problem of the origin of time. It also suggests another kind of Multiverse, if you regard each phase of expansion from a big bang and subsequent collapse to what is sometimes called a big crunch as a separate universe. And with an infinite number of 'past' and 'future' bubbles, there is scope for the kind of variations on the values of the cosmic constants that seem to be required in order to explain the coincidences described in Chapter Two, and therefore our existence. Unfortunately, the bouncing model doesn't work, at least as a description of our Universe. But it is such a fascinating prospect that it is worth explaining why it doesn't work.

Part of the appeal of the cyclic model was that for a long time it was thought that it avoided the need for a singularity at the Big Bang. Taken at face value, Einstein's equations imply that the Universe began from a singularity, a point of infinite density and zero volume. But initially it was thought that a collapsing universe could bounce at some very high density well before it actually reached a singularity. The kind of density people had in mind would be something like the density of an atomic nucleus, the most extreme density of matter in the Universe today. In the 1960s, however, the theorists Roger Penrose and Stephen Hawking showed, in effect by winding the equations back in time, that the singularity at the birth of our Universe could not be avoided within the context of relativity theory, and also that exactly equivalent singularities must lie at the hearts of black holes. The general theory of relativity may not have the last word in this case; what happens close to the singularity depends on how quantum physics and gravity interact. But it is certain that a simple 'bounce' from a state of very high but 'normal' density, such as the density of the nucleus of an atom, is ruled out.

There are also problems with entropy in oscillating models of cosmology. These can be highlighted by thinking about just one phase of expansion and collapse, one bubble in the supposed infinite chain of oscillations. In the 1930s, the American physicist Richard Tolman pointed out that entropy builds up in each oscillation, so that successive 'big bangs' start out with greater entropy. The effect of this is to make successive bounces more vigorous, so that successive oscillations of the universe expand more. If the chain of oscillations really does stretch infinitely far into the past, then the current 'oscillation' should, in fact, be so big that it would be indistinguishable from a unique big bang producing a single expanding universe. This removes the whole point of the oscillatory model. But that hasn't stopped some rather desperate ideas being put forward in an attempt to save it, concentrating on the idea that time might run backwards in the collapsing half of such a universe, thereby resetting the entropy clock.

BACK TO THE FUTURE

In such a situation, the collapsing half of the bubble would be a mirror image of the expanding half, with radiation falling on to stars to 'unmake' complex elements and all the rest that a reversal of the arrow of time would imply. At first sight, this might not seem very different from the idea that the arrow of time points in opposite directions in the two halves of a Boltzmann fluctuation. But there is a crucial difference. In the Boltzmann fluctuation, the arrows are back to back, and both point away from the more ordered state towards the disordered, high entropy state of the greater meta-universe. In the expanding and collapsing bubble universe, the arrows are head to head, and both point towards the state of maximum expansion of the bubble, away from the singularities at either end.

It's like the difference between two trains on the same track leaving a station in opposite directions, and two trains on the same track heading towards each other in opposite directions. What happens where the arrows meet in the middle? We know what would happen to the two trains; in order to avoid something equally messy in a bubble universe, time would have to reverse, instantaneously, everywhere in the universe at the same time. How does everything in the bubble universe know that it is time to reverse all the processes of thermodynamics? To take just a single, simple example, suppose that a single photon of light has just emerged from the surface of a star and is about to set off into the void when the universe reaches its maximum size. How can that single photon (along with every other photon in the universe) suddenly reverse its motion and head back in to the star from which it has just emerged?

Even if we acknowledge that the flow of time is just an illusion, and that the underlying reality is an array of time slices like Fred Hoyle's pigeonholes, the only way this would work would be to change the way the 'memory' stored in each pigeonhole worked at the point in the stack corresponding to the maximum size of the universe. The pigeonholes on the other side of that dividing line would contain 'memories', or records, of what we call the future, not of the past. Both sets of records would describe an expansion from a big bang to a state of maximum expansion. In fact, in order to re-set the entropy clock they would have to describe the same expansion from a big bang. There would be no need for the second half of the stack at all, since it would simply be a duplicate of the first half of the stack. There would not be two big bangs, let alone an infinite array of them, but only one; whichever half of the bubble you lived in, you would see time flowing in just the way we see it today, with a big bang in the past and an expanding universe in the future. We are left with a single big bang that just exists, bringing back with full force the puzzle of the cosmic coincidences. Whatever such an isolated bubble of spacetime might be, it is certainly not a Multiverse, and the idea is hard to take seriously at all.

But there is a remaining puzzle about the arrow of time and the nature of light and other forms of radiation which ought to be cleared up here. I didn't choose the example of a photon leaving a star and then reversing itself arbitrarily, but because the equations that describe the behaviour of electromagnetic radiation work just as well 'forwards' or 'backwards' in time. The equations, discovered by James Clerk Maxwell in the nineteenth century, are usually interpreted to describe radiation which leaves a source, such as the Sun, or a TV transmitter, and moves outward into the universe, where some of it might arrive at your eyes, or at the TV antenna on the roof of your house. Light from a source spreads out across space, like ripples spreading out from a pebble dropped into a pond, but in three dimensions. It's natural for us to interpret Maxwell's equations this way because this interpretation matches our everyday experience of the flow of time. But the equations work equally well in reverse. They also describe what can only be described as electromagnetic waves that start out far away across the Universe and move inwards, converging onto an object like a star. This includes waves that leave the TV antenna on the roof of your house and mingle with waves coming from all directions in the Universe to converge upon the 'transmitter'.

The philosopher Huw Price, who is based at the University of Sydney, has used this kind of symmetry in the laws of physics as the cornerstone of an argument that the single 'bubble' universe with reversing arrow of time and two beginnings but no end has to be taken seriously. His argument is fascinating, and similar ideas were discussed by Stephen Hawking in the 1980s; but they have been rebutted by physicists such as Don Page, of the University of Alberta, and Paul Davies, of Arizona State University, who point out that the correct interpretation of the quantum realities is not that time reverses in any single universe, but that in the many worlds of the Everett interpretation there will be an equal number of universes in which time runs 'backwards' as there are in which time runs 'forwards'. Although a tiny fraction of all the possible worlds could behave in odd ways, which would rule out the possibility of life as we know it, 'a random observer,' says Davies, 'is overwhelmingly likely to find her/himself in a universe with an unwavering arrow of time.'

Even if time does not reverse in the collapsing half of an oscillating universe, and time's arrows continue to point in the same direction, there is another problem with the idea of a bouncing universe. Our Universe emerged in a very smooth state from the Big Bang, and since then gravity has pulled lumps of matter together, ultimately to make black holes, crumpling up the fabric of space. All black holes must contain singularities or, at least, the extreme near-singular conditions that existed at the birth of the Universe, where quantum processes and gravity interact in ways which are still not understood. Matter that falls in to a black hole reaches the equivalent of the big crunch at the end of the universe before the universe itself arrives at the big crunch. So the big crunch itself involves the merger of black holes and singularities, in a complicated process that is definitely not a mirror image of the Big Bang.

Thanks to the repulsive power of dark energy, we are definitely not heading for a big crunch, and none of these speculations about repeating cycles of expansion and collapse apply to our Universe. But that doesn't mean that there may not have been something 'before' the Big Bang or that there may not be something 'after' our Universe fades away. The best explanation we have of how the Universe got started leads naturally into another variation on the theme of the Multiverse, and a satisfactory resolution to the puzzle of the cosmic coincidences. It also involves what we think we know about the behaviour of matter near singularities, and it's called 'inflation'.

5

(Just Like) Starting Over

Cosmologists are confident that they understand in some detail – that is, on the scale of galaxies and stars – everything that has happened in our Universe since it was about one ten-thousandth of a second old. This seems an extraordinary claim to make, but it is based on solid foundations, starting with what we mean by saying that there was a time when the Universe was one ten-thousandth of a second old.

Although the general theory of relativity says that there must have been a singularity at the beginning of the Universe, and that there must be singularities inside black holes, physicists see this as a problem with the general theory rather than a description of the real world. They expect that a quantum theory of gravity will remove singularities from the equations. But if we imagine winding the expansion of the Universe backwards in time in accordance with the equations of the general theory, we can set the moment corresponding to the singularity in the equations as 'time zero'. The question then becomes, how far back can we go – how close can we get to time zero – before we are forced to reconsider what the general theory, and all the other laws of physics that apply today, are telling us?

The most conservative estimate is to start by considering the most extreme form of matter that has been studied in detail here on Earth. Atomic nuclei have been probed by experiments on Earth for well over a hundred years, and their behaviour is thoroughly understood. Our understanding of this behaviour is also applied, with great success, to explain how stars work, with nuclear reactions going on in their interiors. It is safe to say that physicists understand the behaviour of matter at the densities corresponding to the density of an atomic nucleus. From the known density of the Universe today, it is a simple matter to calculate the density at each epoch in the past, when the Universe was smaller but still contained just as much matter as it does today. It doesn't matter if the Universe is infinite, since what we are interested in here is the origin of the part of the Universe we can see today, and how it has changed as time has passed; the same thing may be going on just over our horizon, but that doesn't affect the discussion. The result of this calculation is the discovery that just one ten-thousandth (0.0001) of a second after time zero the entire content of the Universe we see around us today was packed into a hot lump of matter with the density of an atomic nucleus.

This was what is traditionally referred to as the Big Bang; the hot fireball was expanding rapidly at that time, and the known laws of physics can explain entirely satisfactorily how irregularities in that expanding fireball became the seeds on which galaxies grew, with stars and planets forming within those galaxies as the Universe aged, provided allowance is made for the dark matter mentioned in Chapter Two. What they could not explain when the standard Big Bang model was first developed in the 1960s* was where the fireball came from – what happened before 0.0001 sec to set it expanding and to imprint it with irregularities just the right size to grow into the structures we see around us. And why did the Universe have just the critical density of matter?

Forty years ago, at the end of the 1960s, that was as far back as cosmology went. Nobody knew how the Universe had got to be in a state of nuclear density at a high temperature, expanding rapidly away from what seemed to be a singular origin, and even many cosmologists thought that we would never know. But within ten years, by the end of the 1970s, a marriage of particle physics and cosmology had begun to explain what went on before the traditional Big Bang† to set the Universe expanding. And now developments of this idea, which is known as inflation, offer an explanation of the origin of the Universe

---

* The Big Bang *idea* goes back farther than this, but the accurate Big Bang *model* is only a little over 40 years old.
† Today, the term Big Bang is often used to refer to everything from time zero onwards.

itself, within an expanding meta-universe. It may even link us to the Multiverse.

THE PARTICLE CONNECTION

'Particle' physics is something of a misnomer, since, as we have seen, physicists regard fundamental entities such as electrons as quantum phenomena which have properties of both particles and waves. They are described in terms of the behaviour of fields, like the gravitational and magnetic fields. Since forces such as gravity and magnetism are themselves also described in terms of fields, the study of how particles and forces interact is properly called quantum field theory; but particle physics is a less intimidating term, and is widely used even by the quantum field theorists themselves. The important point, though, is that whatever name it goes by, the theory describes the behaviour of particles and forces – the way things like electrons interact with things like electromagnetism, and with other particles.

There are really only two important ideas to take on board. The first is the way that energy can be exchanged between material particles and fields, in line with $E = mc^2$. Since, at the quantum level, this actually involves changing one kind of field into another kind of field, it is less surprising than it might seem at first sight. If there is enough local energy available in a field, it can convert itself into a pair of particles (strictly speaking, a particle and its antiparticle counterpart), and these entities can also interact, disappearing as their energy is converted into another kind of field energy. The simplest example is a high-energy photon (a quantum of the electromagnetic field) that turns into an electron and a positron, in a completely reversible interaction. The second important idea is that all of the forces we find at work in the Universe today – gravity, electromagnetism, and the two forces (dubbed 'strong' and 'weak') that operate only on the scale of atomic nuclei and below – have split off from a single superforce that operates at very high energies.

Physicists have not yet found a unique set of equations to describe this superforce; but the idea of unification of forces is scarcely new. Back in the nineteenth century, the Scot James Clerk Maxwell discovered a set of equations which describe both electricity and magnetism, which had previously been regarded as quite separate forces, in one package, as different aspects of a single force (or single field), electromagnetism. Since then, the quantum version of electromagnetism, known as quantum electrodynamics, or QED, has been combined very satisfactorily with the weak force in a single mathematical package, and there are compelling signs that the strong force can be included in an extended version of this 'electroweak' theory. The big problem is fitting gravity, which is the weakest of the forces, into the same mathematical package as the others. That's why quantum gravity is such a hot topic of research today – as well as providing our best chance of getting rid of singularities, it offers hope of finding a single set of equations to describe all the forces and particles together in what field theorists call the Theory of Everything. More of this in the next chapter; for now, what matters is what particle physics, or field theory if you prefer, can tell us about the origin of the Big Bang.

NOTHING COMES FROM NOTHING

The principal reason why physicists are convinced that quantum gravity will solve the problem of the singularity at the birth of time is that quantum physics tells us that time, like everything else, is quantised. In other words, there is a smallest possible unit of time, which cannot be divided. Certainly, this fundamental unit of time is very small – 10^{-43} sec, which means a decimal point followed by 42 zeroes before you get to the 1. That's why we do not notice the graininess of time in everyday life (although you may relate this graininess to the 'time slices' in the Hoyle/Deutsch description of time). But it is *not* zero. This means that any satisfactory quantum theory of gravity will tell us that the Universe started not from a singularity with infinite density at time zero, but from an extremely dense (but finite) state with an 'age' of 10^{-43} sec, which is known as the Planck time, in honour of the quantum pioneer Max Planck. There was no earlier time. And it turns out that it is very easy to 'make' a universe which starts out with the Planck time as its age; the clever bit is getting it from 10^{-43} sec all the way up to 0.0001 sec.

Surprisingly, this idea, of the Universe as a quantum fluctuation, goes back almost as far as the traditional Big Bang model, although it was laughed out of consideration at the end of the 1960s. It stems directly from the uncertainty relation of quantum physics, which, as I described in Chapter One, tells us that there are pairs of parameters, known as conjugate variables, for which it is impossible to have a precisely determined value of each member of the pair at the same time. This is not, remember, because of any imperfection in our measuring equipment, but is a fundamental feature of the way the world works. One of the most important of these conjugate pairs is energy/time. Quantum uncertainty means that it is impossible for an object such as an electron to have a precise energy – there is always some uncertainty about how much energy it carries. This seems straightforward enough. But quantum uncertainty also tells us that it is impossible for even the energy of empty space to have a definite value, and zero is a precise value.

This means that in any tiny volume of empty space (or, indeed, in a large volume) there might be a bit of energy. This energy can manifest itself in the form of particles, provided the particles are short-lived (this is where time comes into the equation). Instead of a photon converting itself into an electron-positron pair, which then annihilate one another to make another photon, an electron-positron pair can appear out of nothing at all, provided the particles annihilate one another and disappear back into nothing at all within a specific time limit set by the uncertainty relation. Other kinds of particles and forms of energy can appear and disappear in the same way, always within the time constraints set by quantum uncertainty.

According to quantum physics, what we think of as empty space – a vacuum – is actually seething with short-lived entities produced in this way. They are called vacuum fluctuations, or quantum fluctuations. And they aren't just a theoretical prediction – the existence of clouds of these 'virtual' particles around real charged particles such as electrons is necessary to explain the measured strength of electric and magnetic forces. But the length and time scales involved are absolutely tiny. For the example just given, the fluctuation lasts for only about 10^{-21} seconds, and the separation between the electron and the positron never exceeds 10^{-10} cm. The more mass that is involved (which means more energy, of course) the shorter the time for which the fluctuation can exist.

This is all very well when applied to entities such as photons, electrons and positrons. But in the late 1960s the scientific world certainly wasn't ready to consider the idea that the entire Universe might be a vacuum fluctuation when a young researcher called Edward Tryon blurted out the proposal – as much to his own surprise as anyone else's – at a seminar at Columbia University in New York. Tryon had not long completed his PhD work in particle physics, and was very much a junior member of the audience that attended a lecture on cosmology given by Dennis Sciama, visiting from England. During a pause in the presentation, the idea that the Universe might be a vacuum fluctuation popped into his head, and he said it out loud, without thinking. He was so embarrassed by the laughter that followed that he blotted the incident from his mind, and doesn't recall thinking about it at all for several years.

But in the early 1970s some new developments put the idea back on his conscious scientific agenda. Tryon had kept up his cosmological reading, and in 1971, when he was working at Hunter College in New York, he read a review article in the journal *Nature* which discussed the possibility that the Universe might be equivalent to a large black hole, with us inside it.* This idea was picked up by, among others, R. K. Pathria, of the University of Waterloo, Ontario, who put it into a proper mathematical framework in a paper published in *Nature* in December 1972, which Tryon also read. But it was only when a colleague who had also been reading *Nature* reminded him of his outburst at Sciama's seminar that the memory of that occasion came flooding back.

Tryon knew that the total energy of a black hole is zero, because of the trade-off between mass energy and gravitational energy discussed in the previous chapter, and he says that a fully worked-out version of his idea that the Universe might be a quantum fluctuation appeared to him 'suddenly, in a flash' once he was reminded of it. He speculates that his subconscious mind had been working away at the problem for three years, only ready to release it into his conscious

---

* The article, which I wrote, appeared on 13 August 1971.

mind when it was fully worked out and would not provoke more laughter. The fully worked out version of the idea duly appeared in *Nature* in December 1973. It provoked an immediate flurry of interest, but the interest soon waned, because of a crucial problem with the original idea.

Tryon suggested that a quantum fluctuation could occur, on a scale smaller than a proton but containing as much mass-energy as the visible Universe, because all that mass-energy would be cancelled out by the negative gravitational energy associated with the fluctuation. The more energy such a fluctuation has, the shorter its lifetime; but the less energy it has, the longer it can live. Since it would have zero energy, the quantum rules would allow such a universal fluctuation to last forever!

As Lucretius put it, 'nothing can be created from nothing.' For thousands of years, the implication was that the Universe could not have emerged from nothing. Now, Tryon was saying, in effect, that the Universe *is* nothing, turning the aphorism on its head. 'Nothing' with the mass-energy of the Universe, balanced by the negativity of its own gravity, could indeed be created from nothing. But the big snag is that its enormous gravitational pull would make such a superdense 'nothing' collapse into a singularity, regardless of what the rules of quantum uncertainty said. In 1973, nobody knew of any way to inflate such a cosmic seed from the quantum scale up to nuclear densities, all within one ten-thousandth of a second. But at the end of the 1970s, everything changed.

INFLATING THE UNIVERSE

It was only after the standard Big Bang idea was firmly established as a good description of the Universe, at the end of the 1960s, that cosmologists began to worry seriously about the kind of coincidences I discussed in Chapter Two, including the facts that space is very nearly flat with the density of the Universe close to critical, and that the distribution of matter across the Universe is incredibly smooth on the largest scales, but contains irregularities just the right size to allow for the growth of things like galaxies, stars, planets and ourselves.

According to the Big Bang model, these properties were already imprinted on the Universe at the time of the Big Bang, when the Universe was one ten-thousandth of a second old and everywhere was as dense as the nucleus of an atom today. As the evidence mounted that the Universe we see around us has indeed emerged from such a hot fireball, the question of how these properties got imprinted on it became more pressing. They had hardly mattered when people hadn't been sure if there really had been a Big Bang; but now people began to try to work out what the Universe had been like at even earlier times, when it was hotter and denser, in an effort to find out what had set it up in the Big Bang to develop in the way it has.

This quest involved taking on board ideas from high energy particle physics, using theories based on the results of experiments carried out at high energy particle accelerators. These are the experiments and theories that suggest, for example, that entities such as protons and neutrons are actually made up of smaller entities known as quarks, and that the description of all the forces of nature can be combined into one mathematical package. It turned out that in order to understand the Universe on the very largest scales it was first necessary to understand the behaviour of particles and forces (fields) on the very smallest scales and the highest energies.

To put this in perspective, the kind of energies reached by particle accelerators in the 1930s correspond to conditions that existed in the Universe when it was a little over three minutes old; the accelerators of the 1950s could reach energies that existed naturally everywhere in the Universe when it was a few hundred-millionths of a second old; by the end of the 1980s, particle physicists were probing energies that existed when the Universe was about one tenth of a thousand-billionth of a second (10^{-13} sec) old; and the new Large Hadron Collider (LHC) at CERN, near Geneva, is designed to reproduce conditions when the Universe was only 5×10^{-15} of a second old – a fraction of a second indicated by a decimal point followed by 14 zeroes and a 5.

There is no need to go into all the details here, but one crucial point is that the distinction between the four kinds of force that are at work in the Universe today becomes blurred at higher energies. At a certain energy, the distinction between the electromagnetic force and the weak force disappears and they merge into a single electroweak force; at a higher energy still, the distinction between the electro-weak force and the strong force disappears, making what is known as a grand unified force;* and it is speculated that at even higher energies the distinction between these combined forces and gravity disappears.

As far as the early Universe is concerned, higher energies existed at earlier times. So the suggestion is that at the Planck time there was just one superforce, from which first gravity, then the strong force, then the weak force split off as the Universe expanded and cooled. How does that help us? Because, as one young researcher realized at the end of the 1970s, this cooling and splitting off of forces could be associated with a dramatic expansion of the Universe, taking a volume of superdense stuff much smaller than a proton and whooshing it up to the size of a grapefruit in a tiny split second. That grapefruit was the hot fireball, containing everything that has become the entire visible Universe today, that we call the Big Bang.

The researcher was Alan Guth, then (1979) working at MIT, a particle theorist who had become interested in the puzzle of the Big Bang. He realized that there is a kind of field, known as a scalar field, which could have been part of the primordial quantum fluctuation, and would have had a profound effect on the behaviour of the very early Universe. It happens that the pressure produced by a scalar field is negative. This isn't as dramatic as it sounds – it only means that this kind of pressure pulls things together rather than pushing them apart. A stretched elastic band produces a kind of negative pressure, although we usually call it tension. But the negative pressure associ-ated with a scalar field can be very large, and it does have something exotic associated with it – negative gravity, which makes the Uni-verse expand faster (this is essentially the same effect, but on a more dramatic scale, as the lambda field described earlier).

Guth saw that the presence of a scalar field in the very early Universe would make the size of any part of the Universe – any chosen volume of space – double repeatedly, with a characteristic doubling time. This kind of doubling is called exponential growth, and very soon runs away with itself. What Guth did not know at the time, but which

---

* Theories which describe this are called Grand Unified Theories, or GUTs.

made his ideas immediately appealing to cosmologists, was that this kind of exponential expansion is described naturally by one of the simplest solutions to Einstein's equations, a cosmological model known as the de Sitter universe, after the Dutchman Willem de Sitter, who found this solution to Einstein's equations in 1917.

When Guth plugged in the data from Grand Unified Theories, he found that the characteristic doubling time associated with the scalar field ought to be about 10^{-37} sec. This means that in this remarkably short interval any volume of the early Universe doubles in size, then in the next 10^{-37} sec it doubles again, and again in the next 10^{-37} sec, and so on. After three doublings, that patch of the Universe would be eight times its original size, after four doublings 16 times its original size, and so on. After n doublings, it is 2^n times its original size. Such repeated doubling has a dramatic effect. It requires just 60 doublings to take a region of space much smaller than a proton and inflate it to make a volume about the size of a grapefruit, and 60 doublings at one every 10^{-37} sec takes less than 10^{-35} sec to complete.

If we are lucky, the LHC will probe energies that existed when the Universe was 10^{-15} sec old. There may not seem to be much difference between 10^{-15} and 10^{-35}, but that's because we naturally look at the difference between 15 and 35 and think it is 'only' 20; it is actually a factor of 10^{20}; that means that at 10^{-15} sec the Universe was already a hundred billion billion times older than it was at 10^{-35} sec. Putting it another way, the difference between 10^{-15} and 10^{-35} is actually 10^5 times (one hundred thousand times) bigger than the difference between 1 and 10^{-15}. So there is no hope of probing these energies directly in experiments here on Earth – the Universe itself is the test bed for our theories.

This is all based on Guth's original figures. Some modern versions of inflation theory suggest that the process may have been slower, and took as long as 10^{-32} seconds to complete; but that still means that Guth had discovered a way to take a tiny patch of superdense stuff and blow it up into a rapidly expanding fireball.* Even with this more

---

* This rapid expansion seems to proceed faster than light. This is OK, because the speed of light is the ultimate speed that anything can travel *through* space. In inflation, space itself is stretching.

modest version of the expansion, it would be equivalent to taking a tennis ball and inflating it up to the size of the observable Universe now in just 10^{-32} seconds. The process comes to an end when the scalar field 'decays', giving up its energy to produce the heat of the Big Bang fireball and the mass-energy that became all the particles of matter in the Universe. Guth had found the missing link between Tryon's idea of the Universe as a vacuum fluctuation and the traditional Big Bang; the antigravity of the scalar field overwhelms the self-gravity that seemed to be the key problem with Tryon's idea. But at the time, Guth didn't even know about Tryon's work.

The initial, and continuing, appeal of inflation is that it explains many of the cosmic coincidences. As described in Chapter Two, the huge stretching of space involved in 60 or so doublings smooths out irregularities in much the same way that the wrinkly surface of a prune is smoothed out when the prune is put in water and expands. If it doubled in size 60 times (imagine a plum about a thousand times the size of our Solar System) and you stood on its surface you would not even be able to tell that the surface was very slightly curved rather than completely flat, just as for a long time people living on the surface of the Earth thought that it must be flat. In other words, inflation forces the density of the Universe to be indistinguishably close to critical.

The smoothing is imperfect because during inflation 'ordinary' quantum fluctuations will produce tiny ripples which themselves get stretched as inflation continues.* So the distribution of matter in the form of galaxies across the Universe today is only an expanded version of a network of quantum fluctuations from the first split second after time zero. Statistically speaking, the pattern of galaxies on the sky does indeed match the expected pattern for such fluctuations, a powerful piece of evidence in favour of the inflation idea. Many other cosmic coincidences can also be explained within the framework of inflation, since if our entire visible Universe has inflated from a region much smaller than a proton, there may be many other universes that inflated in a similar way but are forever beyond our horizon. And they need

---

* This idea of quantum fluctuations getting stretched in the expanding Universe was first thought of by a Russian cosmologist, Slava Mukhanov.

not all have inflated in the same way – perhaps not even with the same laws of physics.

This has led to new ideas which echo some of the ideas of Boltzmann, and are also reminiscent of another idea which has been, perhaps wrongly, consigned to the dustbin of history – the Steady State model of the Universe.

THE RETURN OF THE STEADY STATE?

In 1980, when Alan Guth was asked how his then new idea of inflation related to the Steady State model of the Universe, he replied, 'What's the Steady State model?' Yet just 20 years earlier, the Steady State model was still regarded as a viable alternative to the Big Bang model. Another quarter of a century down the line from Guth's question, even more water has flowed under the bridge, and even fewer people remember the Steady State model, so it's worth explaining the history in a little more detail before recounting how it relates to the modern theory of inflation.

In the late 1940s, although it had been established that the Universe is expanding, there seemed to be serious problems with the idea that it had come into existence at a definite moment in time when everything we can see around us was piled up at (or near) a singularity. The biggest problem was that at that time estimates of the rate at which the universe was expanding implied that it was only a few billion years old, younger than the oldest stars. This was obviously impossible. There were also lingering philosophical objections to the idea of a Universe with a beginning – a moment when time began. Against this background, three astronomer-mathematicians at the University of Cambridge – Herman Bondi, Thomas Gold and Fred Hoyle – developed the idea that the Universe might be eternal and unchanging in its overall appearance, even though it is expanding at a steady rate.

The essence of the original Steady State idea was that as clusters of galaxies moved apart from one another, the same processes that cause the stretching of space would cause the creation of new atoms of hydrogen in the new space between the clusters, so that the cosmic density of matter always stayed the same. The amount of new matter needed to be created to do the job is only about one hydrogen atom in every 10 billion cubic metres of space every year, which didn't seem terribly extravagant. These atoms would form clouds which over billions of years could condense to form galaxies, stars and all the rest of the material world.

In the 1950s, and even later, to supporters of the Steady State idea, the concept of a steady, continuous creation of hydrogen atoms in this way seemed no more incomprehensible than the idea that all the matter in the Universe suddenly appeared in one go, in a Big Bang. If they were right, then at any moment of cosmic time the Universe would have much the same overall appearance, with, for example, the same number of galaxies and clusters in any particular volume of space, even though they would not always be the same individual galaxies in that volume of space.

This simple version of the Steady State model, based as much as anything on philosophical considerations, is completely wrong. Observations show that as we look farther out across the Universe, which corresponds to looking back farther in time,* we see galaxies that are younger than those near to the Milky Way, and which are closer together than galaxies are today. There is also the evidence from the cosmic background radiation; there is no doubt that the Universe has expanded from a hotter, denser state. Over the decades since the Steady State model was invented, the age problem has also disappeared. Improved measurements of the rate at which the Universe is expanding tell us that the Big Bang happened about 13.7 billion years ago, while improved estimates of the ages of stars tell us that the oldest stars are a little more than 13 billion years old. Everything fits. But the simple version of the Steady State model was not the last word on the subject.

Hoyle in particular, along with his Indian colleague Jayant Narlikar, developed a fully worked-out mathematical version of the Steady State idea, within the context of the equations of the general theory of relativity. The central concept in the model was the idea of a so-called C-field ('C' for Creation) which filled the universe and was responsible

---

* Because of the finite speed of light.

both for the creation of matter and a pressure which caused the universe to expand. In order to make the idea match the ever-improving observations, Hoyle and Narlikar had to retreat from the idea of a steady, uniform creation of matter everywhere in the universe. Instead, they ended up with a version of what they still called the Steady State model in which matter creation is concentrated in what are known as Planck particles with enormous mass-energy contained in a tiny volume, on the quantum scale. The volume is the volume of a sphere with the smallest diameter possible, known as the Planck length. Just as there is no time interval shorter than the Planck time, so there is no distance shorter than the Planck length; the smallest seed from which our Universe can have grown was the size of a Planck particle, the Planck size, with a diameter equal to the Planck length. In round terms, that was one billionth of a trillionth of the size of an atomic nucleus. Yet it contained all of the mass-energy of the Universe we see around us – but overall, once gravity is taken into account, zero energy.

Outbursts from such Planck particles occur, according to the model, within the framework of a larger (infinitely large) meta-universe, creating bubbles of expanding space in a process almost identical to the process I have just described in the context of inflation. The only difference is that in the C-field version the Planck particles are not specifically identified as quantum fluctuations. Hoyle and Narlikar had, in fact, invented a new version of the Big Bang idea, and Hoyle's younger self would probably have been horrified. But because they thought of it in Steady State terms, they did not see what the equations were telling them.

In the 1960s and 1970s, the efforts made by Hoyle and Narlikar to sustain the Steady State idea in the face of improving observational evidence seemed to most cosmologists a little cranky and obsessive. But what they ended up with was a model in which a very dense state of the Universe is driven into a burst of rapid expansion by the energy of a field which then decays into matter particles. That would also be an accurate word-picture of the basic inflationary model. The field responsible for inflation in the standard inflationary models is often denoted by the Greek letter phi (ϕ); in the Hoyle–Narlikar model, a field with exactly the same properties is denoted by the letter C. I well remember a meeting of the Royal Astronomical Society in London in December 1994 where Hoyle, by then in his late 70s and understandably irritated that his work never got a mention when inflation was being discussed, gave a presentation in which he showed that the equations of inflation theory are, indeed, exactly the same as the equations of the final version of the Steady State idea, but with the letter 'C' replaced by the letter 'φ'. 'This,' he said with heavy irony, 'makes all the difference.'

The lesson to be learned, if there is one, is that the truth lies in the equations, and our images of what went on long ago are more aids to our imagination than anything else. It turns out that it was as naive of cosmologists not much more than half a century ago to think of the visible bubble of space that we can see as everything that there is as it was of their predecessors not much more than half a millennium ago to think that the Sun goes round the Earth. Neither the original Big Bang idea nor the original Steady State idea were right, and the best description we now have of how the Universe got to be the way it is involves a hybrid of the two – a Big Bang happening within the context of a greater Steady State. Hoyle was less right than he imagined, but more right than his opponents thought, and his ideas do at least deserve a mention in the context of the modern vision of the inflationary Universe.

But it is Alan Guth who deserves, and gets, the credit for seeing the power of inflation, in the context of the Big Bang idea, as a solution to the puzzle of the cosmic coincidences. He even came up with the name, although he doesn't recall exactly how it happened – 'I do not remember ever trying to invent a name,' he says, 'but my diary shows that by the end of December [1979] I had begun to call it *inflation*.'

This was clearly an idea whose time had come. Apart from the unfashionable work of Hoyle and Narlikar, another version of what we would now call inflation had been developed by Alexei Starobinsky, working at the L. D. Landau Institute in Moscow in the late 1970s. This was a much more complicated model than Guth's, based on a quantum theory of gravity, although it contained the same core idea. But at that time, during the Cold War and before the advent of email and the Internet, Soviet scientists had great difficulty communicating with their colleagues in the West, and news of Starobinsky's work did not spread outside the USSR at the time. Once Guth's version of the idea did spread, however, scientists from what would soon be the former Soviet Union played a large part in its development. Inflation soon grew beyond the confines of the traditional Big Bang model. The key development from Guth's original work in the 1980s was indeed an extension of inflation theory to describe not one Universe, but many universes – bubbles on the river of time.

BUBBLES ON THE RIVER OF TIME

Hardly surprisingly, Edward Tryon soon revived the idea of the Universe as a quantum fluctuation of the vacuum in the context of inflation, and in the early 1980s this was elaborated further by Alex Vilenkin, by then at Tufts University. Vilenkin had been born in Kharkov, part of the then USSR, and graduated from Kharkov State University in 1971, but was unable to move on to study for a PhD because the authorities in the Soviet Union regarded him as 'unco-operative'. After five years making a living by doing various odd jobs (he told me that his favourite was being a night watchman in a zoo) and studying physics in his spare time, he was allowed to emigrate to the United States in 1976. Thanks to the scientific work he had already done, he was able to obtain a PhD from the State University of New York the following year. Vilenkin takes the idea of a quantum fluctuation to its logical extreme. Other researchers, such as Tryon, talk of a 'vacuum fluctuation', which implies that there was a vacuum (some form of spacetime) to fluctuate. But Vilenkin tries to describe mathematically the emergence of space, time and matter out of literally nothing at all. He may or may not succeed. But the most successful versions of inflation so far do not need to go to such extremes, because they are set within the context of an eternal meta-universe, with no beginning of time.

The fact that cosmologists talk about different versions of inflation shows that this idea is still a work in progress. Although the evidence for some form of inflation is compelling, there is no single, definitive inflationary model which can explain all the observed properties of the Universe, but rather a choice of detailed models. From the early days of the development of the idea, as Guth appreciated, there have been problems explaining exactly how inflation got started (what was it that inflated?) and how it ended (what happened to make the scalar field give up its energy in the form of matter?). I won't attempt to go into the subtleties of the different variations on the theme, but will stick with the overall picture and describe one of the most appealing versions of the idea, known as chaotic inflation.

Early versions of inflation suffered from the problem of a need for fine-tuning. To take one of the simplest examples, if inflation had stopped before it had completed the job of flattening space, the resulting universe would be obviously curved, in the sense that space would be bent round on itself, and more lumpy than the Universe we see around us, with greater concentrations of matter. Quite possibly, such a universe would not be suitable for life. Inflation comes in all sizes. So how did the Universe emerge from inflation with just the right properties to be a suitable home for life forms like us? The resolution of the fine-tuning problem is the same as the resolution of all the cosmic coincidences discussed in Chapter Two – but now we have a proper cosmological context to discuss this in, rather than just resorting to philosophical speculations about other worlds. There must be a choice of universes, and the nature of life forms like ourselves selects the kind of universe we see around us.

The idea was taken up and developed in the context of inflation by Andrei Linde, a Moscow-born cosmologist who, unlike Vilenkin, was able to pursue his academic career in the Soviet Union right up to the level of a PhD (awarded in 1974) and beyond. He moved to the United States at the end of the 1980s, and is now at Stanford University; but his idea of chaotic inflation was developed when he was working at the Lebedev Institute in Moscow. Similar ideas to some of Linde's work on inflation in the early 1980s were developed independently by Paul Steinhardt and Andreas Albrecht at the University of Pennsylvania.

Linde's key realization was that there need not be anything unique, or even anything special, about the Planck-sized region of spacetime that first inflated, and then expanded in a more leisurely fashion, to become our visible Universe. There could be other Planck-sized regions, perhaps even infinitely many of them, in a larger, perhaps infinitely large, region of spacetime in which all kinds of scalar fields were, and are, at work. On the quantum scale, what we casually refer to as 'empty space' is actually a seething foam of quantum fluctuations, and inflation could set in, Linde realized, in any of them.* Some of the Planck particles would indeed only expand for the shortest of times before collapsing again, in just the way that critics of Tryon's original idea had suggested. Others might inflate a little, but never enough for them ever to allow interesting things like stars and people to form. Others might expand so fast that matter is spread far too thin for stars and planets ever to form. The overall result would be a chaotic mess of bubbles all expanding at different rates in different regions of spacetime, with chaos once again used in the everyday sense, not in the way the term is used in mathematical chaos theory. An analogy might be made with the bubbles in a freshly opened bottle of champagne – but the bottle would have to be infinitely big for the analogy to begin to do justice to the real thing.

It was Linde who called this version of cosmology 'chaotic inflation'. It immediately offers, in a natural way, what is becoming the standard anthropic explanation of the cosmic coincidences. Out of an infinite array of possible bubble universes floating in the meta-universe, we inevitably find ourselves in one suitable for life because there are no life forms around to notice what is going on in bubbles that are not suitable for life. Other bubbles may not just have different sizes and expansion rates, but may have different values for things like the strength of gravity or the efficiency of nuclear burning, thanks to the way the scalar fields interacted in the Planck particle, and how the fundamental forces split off from one another. Even the number of fundamental forces, and the nature of fundamental particles, may be different in different bubbles. Max Tegmark calls this a 'Type II' Multiverse.

There's a further bonus with chaotic inflation. In the original version of inflation, just as in Tryon's version of the appearance of the Universe from a quantum fluctuation, and in the standard Big Bang model of the 1960s and 1970s, there is the question of what came 'before'

---

* It could even set in in a quantum fluctuation within our own Universe; but this would not blast us out of existence, as I shall explain in Chapter Seven.

the singularity, or near singularity, at the birth of the Universe. Chaotic inflation suggests that our Universe grew out of a quantum fluctuation in some eternal region of spacetime, and that exactly equivalent processes create other regions of inflation elsewhere in the meta-universe. In the meta-universe of chaotic inflation (just as in Hoyle and Narlikar's C-field cosmology) there is no beginning and no end of time. This means that chaotic inflation may be eternal. Unfortunately, there is also a related idea known as 'eternal inflation' (another term invented by Linde), which is not quite the same thing. Chaotic inflation tells us, among other things, how the Universe may have begun; eternal inflation, which I will describe shortly, tells us, among other things, how the Universe may end. I mention it now only to avoid any confusion with 'eternal' chaotic inflation.

Although many details remain to be resolved, and mathematical physicists delight in arguing about, and calculating, the properties of the kinds of scalar fields needed to do the job of inflation, this is very much the current 'best buy' in cosmology. Guth now says that 'we don't know all the details, but the evidence is very convincing that the basic mechanism of inflation is correct.'

Some of the remaining detailed issues the mathematicians love to debate seem as esoteric to the lay person as arguments about the number of angels that can dance on the head of a pin. There is, though, one area where the debate brings in the exciting idea of linking inflation with our understanding of fundamental physics at a new level, involving entities known as strings and membranes. This gives us yet another take on the Multiverse, and will be the subject of the next chapter. But first, there is one more point which has recently been the subject of intense debate among the experts and is essential to round off my discussion here – remembering how implausible the idea of a universe-sized Boltzmann fluctuation seems when it is inspected closely, how likely is it that a universe like our own will emerge from an inflating quantum seed? Is it really true that a universe that is born simple but becomes complex as it ages is more likely than a fluctuation that produces a version of you sitting alone in a room surrounded by eternal chaos?

ETERNAL INFLATION AND
SIMPLE BEGINNINGS

The idea of what is now known as eternal inflation occurred to Alex Vilenkin in 1983. He realized that once inflation starts, it is impossible for it to stop – at least, not everywhere. The most natural thing for the inflation field to do is to decay into other forms of energy and ultimately matter; but within any region of inflating space, thanks to quantum uncertainty there will be variations in the strength of the scalar field, so that in some rare regions it actually gets stronger, and the rate of inflation increases. Within that region itself, the most natural thing for the inflation field to do is to decay into other forms of energy and ultimately matter; but, thanks to quantum uncertainty there will be variations in the strength of the scalar field, so that in some rare regions of that region it actually gets stronger, and the rate of inflation increases. The whole pattern repeats indefinitely, like a fractal.

In a statistical sense, there are very many more places where inflation stops and bubble universes like our own (or unlike our own!) develop; but because inflation generates a lot of space very quickly, the volume occupied by inflating regions greatly exceeds the volume occupied by the bubbles. Although there is a competition between the decay of the scalar field producing bubbles devoid of inflation and rare fluctuations making more inflation, the latter are totally dominant. Vilenkin likens this to the explosive growth of a culture of bacteria provided with a good food supply. Bacteria multiply by dividing in two, so they have a typical growth rate, overall, with a characteristic doubling time – exponential growth, just like inflation. Some bacteria die, when they are attacked by their equivalent of predators. But if the number being killed is less than a critical proportion of the population, the culture will continue to grow exponentially. Within the context of inflation the situation is slightly different – the regions that keep on inflating are rare statistically but overwhelmingly dominant in terms of the volume of the meta-universe they occupy. Because there are always quantum fluctuations there will always be some regions of space that are inflating, and these will always represent the greatest volume of space.

Vilenkin's colleagues were initially unimpressed by his idea, and although he published it, he didn't pursue it very actively in the 1980s and 1990s. One of the few people who took the idea seriously was Andrei Linde, who developed it within the context of his idea of chaotic inflation, published a paper on the subject in 1986, and coined the name eternal inflation. In 1987, using the word 'universe' where I have been using 'meta-universe', he wrote that 'the universe endlessly regenerates itself and there is no global "end of time" . . . the whole process [of inflation] can be considered as an infinite chain reaction of creation and self-reproduction which has no end and which may have no beginning.'* He called this 'the eternally existing chaotic self-reproducing inflationary universe'. The idea still wasn't greeted with much enthusiasm, and although Linde promoted it vigorously eternal inflation only really began to be taken seriously in the early years of the twenty-first century, after the discovery of evidence for dark energy and the acceleration of the expansion of the Universe.

All the evidence now points to the likelihood that our Universe will keep on expanding forever, at an accelerating rate. The process is exactly like a slower version of the inflation that produced the bubble of space we live in. Eventually – and it doesn't matter how long it takes since we have eternity to play with – all the stars will die and all the matter of the Universe will either decay into radiation or be swallowed up in black holes. But even black holes do not last forever. Thanks to quantum processes, energy leaks away from black holes in the form of radiation. This happens at an accelerating rate, and eventually they disappear in a puff of gamma rays. So the ultimate fate of our Universe is to become an exponentially expanding region of space filled with a low density of radiation. This is exactly the situation described by the solution to Einstein's equations found by de Sitter, and known as de Sitter space.

De Sitter space is the perfect breeding ground for inflation. Within de Sitter space, quantum fluctuations of the traces of radiation and the scalar field we call dark energy will produce a few, rare Planck-scale

* See *300 Years of Gravitation*, ed. S. Hawking & W. Israel, Cambridge UP, 1987, p. 618.

regions that inflate dramatically to grow into bubble universes like our own. And once you have inflation, then, as Vilenkin and Linde told us more than two decades ago, you have eternal inflation. The discovery of the universal acceleration and its associated cosmological constant link us both with the future and the past of eternal inflation, possibly in a chaotic meta-universe. It suggests that our Universe was born out of de Sitter space, and will end up as de Sitter space. It's just like starting over – and over, and over, and over again. If you do the calculations correctly, it also resolves the puzzle of why small fluctuations that inflate to become complex universes are more likely than fluctuations which produce solitary people sitting alone in a room surrounded by chaos.

BOLTZMANN'S BRAIN, THE ARROW OF TIME, AND CAUSAL PATCH PHYSICS

Even in the context of inflation and exponentially expanding de Sitter space, the thermodynamic arrow of time is still relevant to the existence of our own bubble Universe, even if the greater meta-universe is essentially timeless. The fluctuations that make universes are still fluctuations away from equilibrium, and the way they return to equilibrium is what gives us our arrow of time. The Universe we live in did start with low entropy, and it will end with high entropy. The ultimate fate of exponentially expanding space, containing nothing except the dark energy responsible for the cosmological constant, is in effect the opposite of a black hole. A black hole is the ultimate state of collapse; de Sitter space is the ultimate state of uncollapse. Using the equations of black hole physics in a sense turned inside out, cosmologists can calculate the entropy of de Sitter space, and it turns out that it is, indeed, very high – as high as it is possible for the entropy of something that starts out like our Universe and has a cosmological constant to be. So far, so good. The Universe is going in the right direction, thermodynamically speaking, given where it started from. Which takes us back to the question that puzzled Boltzmann – how did it start in a low entropy state?

In the early years of the present century, a team of researchers from Stanford University and MIT* put the cat among the cosmo-logical pigeons by suggesting that even within the context of inflation, the overwhelming majority of states which could have evolved into a world similar to ours would not start from a low entropy state. The puzzle suggested that it is still easier to make a single individual sitting in a room, or a naked brain complete with (false) memories of learning about the Big Bang and the history of the Universe (and equally false memories of having read about Boltzmann fluctuations earlier in this book†), than to make the Universe itself. This is some-times known as the 'Boltzmann's brain' paradox, since a naked brain lasting for just long enough to 'know' all the things we think we know about the Universe seems to be the simplest statistical fluctu-ation that would explain why you think you are sitting there reading this book.

The argument was based on applying the appropriate equations to, in effect, the region of spacetime that makes our visible Universe. Since nothing can travel faster than light, only objects in regions of space that are close enough together for light signals to have crossed between them since the Big Bang can affect one another. One such object can cause a change in another object in the same patch of the Universe, but it cannot cause anything to happen outside that patch. For obvious reasons, such a region of spacetime is called a causal patch. In causal patch cosmology, physical meaning is only given to events which are said to be causally connected. This is equivalent to treating the Universe as a single, finite system evolving towards a true equilibrium state, de Sitter space, and this was the approach used by the Stanford/MIT team.

But the whole point of ideas such as chaotic inflation and eternal inflation is that there is a much larger (perhaps infinitely large) region of spacetime beyond the boundaries of our bubble – beyond the cosmic horizon. Even though there is no way that anything we do can ever influence what happens outside our bubble, and no way that anything happening outside our bubble can influence us, it turns out that it is

---

* Lisa Dyson, Matthew Kleban and Leonard Susskind.
† Actually, of course, if you are simply a naked brain, the *memories* are real, but the events you remember never happened.

possible to analyse the statistical implications of the existence of such a meta-universe. This is equivalent to treating the Universe as a tiny part of a much larger system which spends by far the majority of its time in equilibrium.

The key difference this makes to the analysis is that within the larger meta-universe fluctuations can go in either direction, towards lower entropy or towards higher entropy, and that the entropy goes up just as often as it goes down. When Andreas Albrecht and Lorenzo Sorbo, then of the University of California, Davis, carried the appropriate calculations through in 2004, they found that not only is inflation overwhelmingly the most likely kind of fluctuation to occur – overwhelmingly more likely even than a fluctuation that produces a 'standard' Big Bang – but even a fluctuation producing the standard Big Bang is overwhelmingly more likely than a fluctuation which produces a naked brain. The essential point that makes the difference is that the small entropy of the fluctuation that will eventually grow to become a universe like our own provides a kind of credit which allows for more of the rest of the meta-universe to remain in equilibrium.

Albrecht and Sorbo make an analogy with a large box uniformly filled with radiation that is in equilibrium on average, but in which statistical fluctuations occur. Imagine a situation in which all the radiation in a volume of one cubic centimetre in one corner of the box shrinks into a volume of one cubic millimetre. This increases the entropy in that cubic millimetre of the box. Now imagine a similar fluctuation in which all the radiation in a volume two cubic centimetres across shrinks into a volume of one cubic millimetre. Once again, entropy increases in that cubic millimetre of the box. In the second example, the entropy of the region one cubic millimetre in size is bigger than that in the first example, so if you only look at the local scene thermodynamics would say that it is more likely. But if you look at the whole box, you see that in the first example more of the box of radiation is in equilibrium, so the *first* fluctuation is more likely than the second one even though it leads to a state which locally has less entropy. For the box as a whole, this state has *more* entropy than in the second example. The difference, which is borne out by the mathematics, results from a difference in perspective, which is like the difference between applying causal patch physics and looking at the meta-universe beyond our cosmic horizon.

The key point is that inflation only requires the smallest possible volume, a quantum of volume, to get started, so it leaves all the rest of the meta-universe in equilibrium, at high entropy. What's more, that quantum is in a very simple state. But to make a universe like our own causal patch, within the uniform meta-universe, *without* the help of inflation you would need a fluctuation as big as a grapefruit, made up of an unimaginably large number of quantum volumes arranged in a complicated way. This is vastly less likely.

The surprising, but intriguing implication is that although nothing outside our causal patch can ever affect us, we are only here because everything outside our causal patch exists. Since the assumption that the Universe is a single, finite system leads us directly to the Boltzmann brain paradox, while the assumption that the Universe is one of many bubble universes in an infinite meta-universe leads to the conclusion that such universes are very likely to exist, the very fact that we exist seems to be the best evidence available that we do not live in a finite, single Universe. And the everything beyond our causal patch, as Vilenkin emphasizes, includes many, many copies of ourselves. Instead of you existing (briefly) as an isolated Boltzmann brain, the meta-universe contains infinitely many copies of your brain, plus infinitely many copies of everybody else's brains, all housed comfortably in living bodies.

TO INFINITY – AND BEYOND!

In the movie *Toy Story*, the character Buzz Lightyear has the catchphrase 'to infinity – and beyond!' Buzz isn't presented as the sharpest pencil in the box, and the scriptwriters' little joke is, of course, that there is no 'beyond' where infinity is concerned. But maybe the joke is on them. According to Alex Vilenkin, our Universe is infinite, but there is indeed a 'beyond' within which our Universe is just a small part. This is a big step beyond the idea of our Universe as a large but finite bubble in an infinitely large meta-universe, and it results from the different points of view that are permissible (indeed, obligatory) within the framework of the general theory of relativity.

The idea of our Universe as a large but finite bubble in an infinite eternally inflating meta-universe corresponds to what Vilenkin calls the 'global view' of the cosmos, in which there is no beginning of time and no obvious way to match up the passage of time measured in each bubble universe with what is going on outside. Indeed, from a thermodynamic perspective it's not clear that the passage of time means anything in the meta-universe, which always has the same overall appearance, on average. But within each bubble universe, any intelligent observers will perceive a definite beginning of time, corresponding to a big bang 'origin' for their own universe. The curious result is that from the internal viewpoint of such observers each bubble universe appears to be infinite in extent.

Vilenkin says that the best way to see this, without going through all the mathematics, is to imagine counting galaxies. From the global perspective, new galaxies are constantly being born near the boundaries of an expanding bubble universe, and with infinite time to play with this means that there must be an infinite number of galaxies in each bubble. There must be just as many galaxies in the bubble from the internal point of view, even if it is not technically possible for a single observer to see all of them. But from the internal point of view – the kind of view we get from Earth in a Universe 13.7 billion years old – all of these galaxies exist at the same time, and there has only been a finite time in which they could have been produced, since the relevant big bang. The only way to have an infinite number of galaxies form in a finite time is for it to occur in infinite space. So each bubble universe is, from the point of view of its inhabitants, infinite in size. It is said to be 'internally infinite'.

In effect, the change of viewpoint means that the infinite time available from the global perspective is converted into infinite space in the internal bubble perspective. This trading of space for time depending on the viewpoint of the observer is exactly the kind of thing that Einstein discovered with his first theory of relativity, the special theory, more than a hundred years ago. It is a bedrock of modern science, something that physicists are completely at home with, and the full implications have to be taken seriously.

The story so far is amazing enough. But Vilenkin and his colleagues have gone further, analysing the implications for any particular bubble universe, including our own. Indeed, for now we can ignore the world outside our Universe and just look at the implications for the Universe we live in.

All the evidence presented in this chapter tells us that our Universe is an infinitely large bubble of expanding space – at least, it is from our perspective, which is what matters. Any particular observer in the Universe – such as an astronomer sitting at an observatory on a mountain top on Earth – can only see as far across the Universe as the distance light has been able to travel since the Big Bang. As we have seen, although the Big Bang happened 13.7 billion years ago, the farthest objects we can see today are actually not just 13.7 billion light years away, but roughly 40 billion light years away, because of the way space has stretched while the light has been on its journey to us. Vilenkin calls the region of the Universe visible to any particular observer an O-region ('O' for 'observable'), and our O-region is a volume about 80 billion light years across. At this moment of cosmic time, when the Universe is 13.7 billion years old, all observers in our Universe are in the same situation. Each of them sits at the centre of an O-region 80 billion light years across. Since the Universe is infinite, this means that there is an infinite number of O-regions. You might think that they are all different from one another; but Vilenkin thinks otherwise.

He gives an example in which one person – you, perhaps – shifts their chair back by one centimetre. There will be another O-region identical to ours in every respect except that the chair has not been moved, another in which the chair is moved by 0.9 cm instead of 1 cm, another in which the chair is moved by 0.99 cm, another in which it moves by 0.999 cm, and so on. It looks like an infinite sequence, in which there are infinitely many possible differences in the final position of the chair. But Vilenkin points out that because of quantum uncertainty you come to a point where the differences in the positions of the chair cannot be detected – not just because they are too small to measure, but because there is no meaning to such small distances.

This means that there is only a finite number of possible positions for the chair – a mind-bogglingly large number, but still finite. By the same token, applying this reasoning to everything that is present in an O-region, from sub-atomic particles to superclusters of galaxies, there is only a finite number of ways in which O-regions can be distinguished from one another. And in an infinite Universe, this means that there must be an infinite number of O-regions that are exactly the same as each other. In the opposite of the scenario of a single Boltzmann brain floating (temporarily) in chaos, there is an infinite number of copies of you, reading an infinite number of copies of this book, on an infinity of Earths in every way identical to our own. As Vilenkin puts it, 'there should be an infinity of regions with histories absolutely identical to ours.'

There are also, just as in the Many Worlds Interpretation of quantum theory, vast numbers of O-regions where things are slightly different, and even more O-regions in which things are very different from our own. 'There are many more ways for things to be different,' says Vilenkin, 'than for them to be the same.' But the important point about all this is that it does not require the Many Worlds Interpretation; it all happens in one single infinite Universe made up of three dimensions of space plus one of time. Of course, the Many Worlds Interpretation may apply as well, but as Vilenkin puts it, his view of the Universe gives you 'many worlds in one'.

All of this puts mathematical flesh on the bones of the philosophical speculation that in an infinite Universe anything can happen – as the Russian scientist and activist Andreas Sakharov put it in 1975, when accepting his Nobel Peace Prize:

In infinite space many civilizations are bound to exist, among them societies that may be wiser and more 'successful' than ours. I support the cosmological hypothesis which states that the development of the universe is repeated in its basic characteristics an infinite number of times.

Eternal inflation elevates these ideas to something more than a mere hypothesis, by bringing them firmly into the realm of physics. It ensures that universes like our own are internally infinite, and quantum fluctuations during inflation ensure that random processes generate all possible variations on the universal theme, rather than, for example, an infinite universe filled with an infinite number of identical copies of our Milky Way galaxy. Cosmologist John Barrow, of the

University of Cambridge, emphasizes the importance of this new understanding of the Universe. Eternal inflation is, he says,

tantalizingly close to providing a sweep of the collection of all logically possible alternatives for the structure of universes within a very wide range of possible worlds.

These examples are important to our understanding of the idea of all possible worlds because they show how the laws of physics can produce a realization of an infinite range of logically possible conditions within our Universe, if it is infinite in size, *without us having to appeal to metaphysical notions like 'other' universes existing in parallel realities*. One infinite universe contains enough room to contain all these possibilities. This is the conservative multiverse option.*

Or as Alan Guth puts it, 'inflation pretty much forces the idea of multiple universes upon us.'

You might think that this is the last word, and that the search for the Multiverse is over. In a sense it is – we don't *need* anything more than the existence of the Universe we see around us, and the ideas presented in this chapter, to justify (or require) the idea of the Multiverse. But theoretical physicists seldom stop with what they *need* to know. As Barrow says, to stop the search there would be the conservative option. The revolutionary option is to embrace new ideas that even go beyond those of the Many Worlds Interpretation, based on developments in 'particle' physics which suggest that even our four-dimensional spacetime is just some kind of 'local' feature within a multi-dimensional framework. According to these speculations, the string's the thing, and there is yet another way to trigger inflation, or something very like inflation.

* *The Infinite Book*, p197; his emphasis.

6

The String's the Thing

The modern understanding of the physical world says that instead of regarding fundamental entities such as electrons as tiny spheres or mathematical points, we should think of them as loops of vibrating stuff, prosaically named 'string'. One way to picture these strings is like tiny stretched elastic bands, vibrating in different ways like the different notes you can play on a single guitar string. One vibration – one 'note' – would correspond to an electron, another mode of vibration might correspond to a photon, and so on. The same kind of fundamental entity, a single kind of string, could be responsible for the appearance of all the different kinds of particles our world is made of, including the particles, such as photons, that carry forces, as well as what we think of as material particles.

This explains the appeal of string theory. It combines the description of all the particles and forces of nature into one package, and seems to be the long-sought Theory of Everything. But it didn't start out that way.

At the end of the 1960s, two physicists working at CERN, Gabriel Veneziano and Mahiko Suzuki, came up with a mathematical description of what happens when particles collide at high energies, like the particles in the beams investigated at CERN. They each noticed, independently, that a particular kind of mathematical expression, called the Euler beta function, could be used in the description of this process. At the time, they had no idea what this might mean in terms of the physical nature of the world of the very small; it was just a useful piece of mathematics, which had been known about in the nineteenth century, that seemed to fit the bill. But when the news spread to America, two young physicists working in different cities each realized that the mathematical expression that seemed to work so well as a description of what happened when particles collide could be interpreted as a description of two loops of 'string' coming together, joining, vibrating for a time and then breaking apart into new loops of vibrating string.

One of these physicists, Yoichiro Nambu, worked at the University of Chicago. The other, Leonard Susskind, was at Yeshiva University in New York. Susskind recalls that at the time he was not working on particle physics at all, but realized that the problems of particle physics could be solved by 'nice, simple mathematics'. He called the vibrating loops 'rubber bands', and for two days he thought he was the only person in the world who knew how particle collisions worked. Then he heard that Nambu had had exactly the same idea at almost exactly the same time,* and hurried to write up his discovery and get it published for the world to see. He needn't have bothered to hurry. The response of the physics community to the early string theory of Nambu and Susskind was almost entirely one of indifference. In the 1970s, the idea that 'particles' such as protons and neutrons are actually made of different combinations of fundamental entities called quarks was hot stuff, and there didn't seem to be any need for strings.

There was, though, one embarrassing feature of the standard model of particle physics based on the idea of electrons, quarks and other entities as point-like particles. If the particles literally existed as mathematical points with zero spatial size (zero dimension), the theories were plagued by unwelcome infinities. In a simple example, think of the inverse square law of the electric force. The force between two charged particles is proportional to a 1 divided by the square of the distance between them. If a charged particle had zero volume, then the force acting on the particle itself (its self-interaction) would be infinite: 1 divided by zero squared. Things like electrons ought to explode as a result.

Theories that include infinities are usually regarded with deep suspicion by physicists, and rightly so; but the standard model was so successful in other ways that they learned to live with it. Ways were

---

* It later turned out that the Danish physicist Holger Nielsen also came up with the idea in a slightly different form, but never developed it fully.

found to fudge round the infinities, but their presence still made the theorists who ventured to think about such things uncomfortable. The standard way to get rid of infinities in quantum field theory, called renormalization, essentially involves dividing one infinity by another so that they cancel out, and any mathematician will tell you that this is a very dubious practice. One of the great appealing features of string theory was, from the outset, that it involved entities with a finite size, so both the zeroes and the infinities were never part of this kind of calculation. At the same time, however, the early versions of string theory also had flaws, and could not be applied to all the known varieties of particle. So the search for a Theory of Everything focused on the development of a description of the world based on quarks, and string theory languished, studied only by a few mathematicians, notably John Schwarz, at Princeton, and Michael Green, in London. But string theory started to be taken seriously when it turned out, almost by accident, that it included a description of gravity.*

GRAVITY GRABS ATTENTION

Including gravity in a Theory of Everything was the Holy Grail of theoretical physicists in the late twentieth century. All the other forces and particles could be included more or less satisfactorily in Grand Unified Theories incorporating, among other things, the idea of quarks; but getting the mathematical description of the physical world to include gravity was tough. There are two reasons. First, gravity is by far the weakest of the fundamental forces; secondly, it has some unusual properties, related to the nature of the particle that carries the gravitational force, called the graviton – the equivalent for gravity of the photon for electromagnetism.

The thing that makes the graviton so difficult to deal with, mathematically, compared with the photon, is that gravitons can interact with each other. Photons interact with charged particles, such as

---

* I'm a little uncomfortable talking about string 'theory' rather than string 'models', because these ideas have not yet passed any experimental test. But it's their success in 'predicting' the existence of gravity that makes them respectable theories to many physicists, so I'll go along with this common usage.

electrons, and carry the electric force from one charged particle to another; but one photon does not directly affect another photon. Gravitons carry the gravitational force from one object to another; but one graviton can directly affect another graviton, and this makes things a lot more complicated. Undaunted by this, in the 1970s theorists tried to find a way to describe the behaviour of gravitons, to complete their picture of the quantum world. In technical terms, what they were looking for, in the context of quantum field theory, was a mathematical description of a massless, spin-2 boson. They had to have this in order to have a complete quantum description of all the particles and all the forces of nature.

While many theorists struggled with this problem, the few string theorists made a double breakthrough. First, they refined the basic idea of strings so that it involved much smaller entities than the 'rubber bands' proposed by Nambu and Susskind. These new kinds of string loops were so small, in fact, that their vibrations could even account for the properties of quarks, the supposedly fundamental entities lurking inside particles such as protons and neutrons. This meant that string theory could reproduce all of the successes of quark theory and its derivatives, *without* the embarrassment of unwelcome infinities. The second breakthrough was one of those 'D'oh – why didn't I think of it sooner?' moments.

TWO APPROACHES PLUS A THIRD WAY

During their efforts to refine string theory to include a description of 'particles' such as quarks, the theorists kept finding that the equations · included a description of a particle they weren't looking for, and which seemed to be getting in the way of their calculations. They tried every way they knew to get rid of it, but failed. It seemed that if they wanted to use string theory to describe the particles they were interested in, they had to include a description of another kind of particle altogether – a particularly complicated kind of particle. It was, as you may have guessed, a massless, spin-2 boson. Without being asked, string theory had produced a mathematical description of the graviton. This breakthrough happened in the mid 1980s, boosted by the discovery, made by Schwarz and Green, of a complete string theory, free from infinities and other anomalies, that was a real candidate for the Theory of Everything. The fact that this theory required the existence of gravitons was, as John Schwarz called it, a 'Deep Truth'. In other quantum field theories, you start with a description of everything else then try to add in gravity; with strings, it is impossible to build a theory without gravity. From then on, people took string theory seriously.

Just how deep a truth the link between string theory and gravitons is can be seen by looking at gravitons from both sides. The two great theories of twentieth-century physics are the general theory of relativity and quantum theory. One aspect of the Theory of Everything – indeed, it's key virtue – is that it would have to combine these two great theories in one package to provide a quantum description of gravity. Starting out from Einstein's description of gravity in terms of curved spacetime, you are led inevitably to the idea of gravitational radiation – ripples in the fabric of spacetime. Quantum theory then tells you that these waves are associated with a particle, the graviton, just as light waves are associated with the photon. But as Richard Feynman spelled out in the 1960s, in his *Lectures on Gravity*,* if you start out with a quantum field theory based on a massless, spin-2 boson you can work upwards to the general theory of relativity. String theory offers the hope of finding the Theory of Everything by starting out from the quantum world, because it gives theorists a reason to start out with a massless, spin-2 boson rather than some other kind of particle.

There is, nevertheless, a second way to approach the Theory of Everything, which starts out from the general theory of relativity and works towards (hopefully) the same conclusion; this is called quantum loop gravity. I won't say any more about it here, because it adds nothing to the story of the search for the Multiverse, but it is worth being aware that string theory is not the only game in town – even though one of the leading proponents of quantum loop gravity, Lee Smolin, has acknowledged that 'it cannot be over-emphasized that in the language in which it is understood – that of diagrams correspond-

---

* Addison-Wesley, Reading, Mass., 1964.

ing to quantum particles moving against a background spacetime – string theory is the only known way of consistently unifying gravity with quantum theory and the other forces of nature.'

Actually, there is a third way, as well: a few brave theorists try to derive the Theory of Everything from first principles, ambitiously anticipating that this will then lead to new ways of understanding both relativity theory and quantum physics. It is a heroic endeavour, but not one to discuss here.

When more physicists started to take the ideas of string theory seriously, from the mid 1980s onwards, they had to confront a particular oddity of this way of describing the quantum world. We perceive the world as occupying four dimensions – three of space plus one of time. But in every variation on the string theme, the equations only work if the strings occupy a world with many more dimensions, at least 11 in all – ten of space plus one of time. This looks alarming at first sight, but it turns out to be nothing new. The idea of unifying the forces of nature by invoking extra dimensions goes back to the 1920s, just after the discovery of the general theory of relativity, and ever since then there has been a standard way to hide the unseen extra dimensions. It is called 'compactification'.

COMPACT BUT PERFECTLY FORMED

When Einstein came up with the general theory of relativity, in the second decade of the twentieth century, the only two forces of nature that physicists knew were gravity and electromagnetism. The general theory was such a success that it was natural for some people to try to apply the same approach that Einstein had used for gravity to electromagnetism. One person who did so, with considerable success, was Theodor Kaluza, a junior researcher at the University of Königsberg. The seed of his breakthrough idea occurred to him early in 1919, but it wasn't published until 1921; five years later, it was refined by the Swedish physicist Oskar Klein, and the complete package is known today as 'Kaluza–Klein theory'.

It's unlikely that anybody would have set out to find a new way of describing electromagnetism in 1919, because there was already a perfectly good set of equations, discovered in the nineteenth century and named Maxwell's equations after their discoverer, the Scot James Clerk Maxwell, that provided a complete description of all interactions involving electricity and magnetism. They later had to be modified to take account of quantum effects, producing the even more successful theory of quantum electrodynamics,* but if quantum effects were left out of the calculation, in 1919 there was a perfectly good 'classical' theory of electromagnetism, provided by Maxwell, and a perfectly good 'classical' theory of gravity, provided by Einstein. Kaluza was just playing with Einstein's equations, the way mathematicians do, especially with new discoveries. Einstein's equations describe gravity in terms of distorted four-dimensional spacetime. Kaluza wrote down the equivalent equations in five dimensions, to see what they would look like. In a moment of revelation, he saw that what they looked like were the field equations of the general theory of relativity, plus another set of field equations mathematically identical to Maxwell's equations.

If gravity could be described in terms of ripples in four-dimensional spacetime, electromagnetism could be described in terms of ripples in five-dimensional spacetime. At a stroke, Kaluza had unified the two classical theories of physics in one mathematical package. Even Einstein was impressed. He wrote to Kaluza that 'I like your theory enormously,' and after discussing some points of detail with Kaluza recommended the publication of the idea in the *Proceedings* of the Berlin Academy of Science.

Even with Einstein's endorsement, the original Kaluza–Klein theory didn't catch on, for two reasons. The first was that as physicists began to discover more particles and forces of nature, which interact in more complicated ways than interactions involving electromagnetism, it no longer stood up as a Theory of Everything, but was 'only' a theory of gravity and electromagnetism. The second was the puzzle of that extra dimension of space required by the theory. Where is the mysterious fifth dimension?

---

* Klein's achievement was to include some of the requirements of quantum theory into Kaluza's model. He rewrote a mathematical description of the electron discovered by the Austrian Erwin Schrödinger, and known as the Schrödinger equation, in five dimensions.

In fact, there is a perfectly straightforward answer to that question, but not one which many physicists found appealing in the decades following Kaluza's insight. The extra dimension could be rolled up very small, or compactified, so that we cannot see it.

The usual analogy is with a drinking straw. If you take an ordinary drinking straw and look through it like a telescope, it is easy to see that it is made of a flat sheet of two-dimensional material* wrapped around the third dimension to make a hollow tube. But if you place the same drinking straw on the floor and look at it from far away, all you see is a line, which looks from a distance like a one-dimensional object. Every 'point' on the 'line' is really a tiny circle, too small to see. In the original Kaluza–Klein theory every 'point' of space is really a tiny loop, no more than 10^{-32} cm across, bent round a direction that is neither up-down, left-right, or forward-back; it is the fourth dimension of space, or the fifth dimension of spacetime.

That might just about have been acceptable to mathematical physicists in the early 1920s. But as more particles and forces were discovered, the need for extra dimensions increased. In the end, to describe all the particles and forces of nature, it turned out that modern versions of Kaluza–Klein theory require at least eleven dimensions, ten of space plus one of time. Which means that no less than seven of the space dimensions have to be rolled up, or compactified, on the quantum scale to leave our four familiar dimensions as the stage on which events take place on the large scale. Only a few mathematically inclined physicists had bothered to keep up to date with Kaluza–Klein theory as the decades passed; but when string theory also came up with the requirement of a minimum of eleven dimensions for the strings to be vibrating in, things changed. Compactification became a key component of the new candidates for the Theory of Everything that emerged in the 1990s and were developed in the early years of the twenty-first century.

* We ignore the thickness of the material itself.

THE MAGIC OF M

The most worrying aspect of string theory was that, in spite of its success, there was no unique version of the theory. There were, indeed, many stringy candidates for the Theory of Everything. To be precise, by the early 1990s the theorists had come up with five different versions of string theory, each of which made mathematical sense and seemed a suitable candidate for the Theory of Everything, and each involving six compactified dimensions plus the usual four dimensions of our familiar spacetime. There was also a sixth theory, known as supergravity, which required eleven dimensions. But the good news was that it was possible to prove that these were the only workable string models – there might be other mathematical packages involving strings, but they would all be plagued by infinities and the problem of renormalization.

But this was small comfort. Too many string theories was almost as embarrassing as too few. As John Barrow has put it: 'You wait nearly a century for a Theory of Everything then, suddenly, five come along all at once.' The best hope was that one of the candidates would turn out to be better than all the others. But in 1995 Ed Witten, at Princeton, reached the opposite conclusion. They were all equally good, he said (and that included eleven-dimensional supergravity) because they were all part of the same thing. He showed that there are ways to transform each of the six models into each of the others, implying that they are different facets of some underlying single theory, the true Theory of Everything. This is just like the way that electricity and magnetism seem to be two separate forces, but are actually different facets of electromagnetism. Another way of looking at it is to imagine that in the days of sailing ships five different European expeditions had separately discovered Alaska, Florida, Panama, Brazil and Tierra del Fuego, and everybody in Europe thought they were five separate islands, not realizing that they were all part of a single connected land mass.

Among other things, the discovery that all these models were different aspects of a single underlying theory meant that the string models, like supergravity, actually required ten dimensions of space, plus one of time. But this time, the 'extra' dimension – the eleventh dimension – would not have to be compactified. It can be very big, but it lies in a *spatial* dimension (not to be confused with the time dimension!) at right angles to all of the other familiar dimensions of space. If our entire Universe were represented as a flat sheet of two-dimensional paper lying on a table, this extra dimension would be at right angles to the surface of the paper, extending upwards in the third dimension.

All of this makes a profound difference to string theory. Instead of thinking in terms of vibrating strings, we have to think in terms of vibrating sheets, or membranes, like the skin of a drum. A membrane is, strictly speaking, a two-dimensional sheet. In mathematical language, a point is a o-brane, a line (or string) is a 1-brane, a sheet is a 2-brane, and, although they are hard to visualize, there are equivalent structures in higher dimensions, known as 3-branes, 4-branes and so on. These are often referred to generically as '*p*-branes', where *p* can be any number. Witten dubbed the whole package of ideas 'M-theory', but has never said what the 'M' stands for; many people are happy to read it as short for 'membrane'.

Membranes can look like strings because of compactification – a particularly simple kind of compactification. Instead of thinking of a one-dimensional line of stuff making up a string, we ought now to be thinking of a ribbon of stuff with a finite width, actually rather like a real rubber band, which is made of a strip of elastic stuff. If the dimension corresponding to the width of the rubber band shrinks until it is too small to see, the ribbon of material will look like a string, even though it is still really a ribbon.

M-theory is the best candidate yet for the Theory of Everything. It also provides new insights into the search for the Multiverse, in more ways than one. The most obvious relates to the image of our entire Universe as a flat sheet of two-dimensional paper lying on a table, with an extra dimension at right angles to the surface of the paper, extending upwards in the third dimension. There is no reason why there couldn't be another sheet of paper on top of the first one, and another, and another – or a multitude of three-dimensional universes separated from one another in the eleventh dimension. This is quite a different concept from the idea of a multitude of bubble universes separated from one another by vast distances across three-dimensional space – the 'universe next door' could be separated from us by only a tiny distance in the eleventh dimension – and I'll come back to it shortly. But M-theory also suggests how different Big Bang universes might follow one another as a sequence in time, with one universe being born, Phoenix like, out of the ashes of a previous incarnation. It has to do, curiously, with the incredible weakness of gravity, and the existence of open-ended strings.

REVISITING THE INCREDIBLE WEAKNESS OF GRAVITY

It's hard to get a grip on just how weak gravity is compared with the other forces of nature. This is so important that before looking at the concept of the Phoenix universe it's worth revisiting some of the ideas mentioned in Chapter Two, with a slightly different perspective. We are used to thinking of gravity as a strong force in everyday life, holding us down on the surface of the Earth and giving us weight. But it takes the gravity of the entire mass of the whole planet to pull you down with your weight, or, as Isaac Newton noticed, to make an apple fall from a tree. Even a small child can pick an apple up off the ground, overcoming the pull of the planet with the muscle power in its arm – muscle power that is, essentially, an aspect of the electromagnetic force, which is vastly stronger than the gravitational force.

In everyday life, we don't notice how weak gravity really is because it has two important properties. First, it is long range – the gravitational force with which an object tugs on another object does fall off as 1 divided by the square of the distance between the two objects (the inverse square law), but in principle it extends forever, however feebly. So the Earth keeps us stuck to its surface, the Sun keeps the planets held in orbit around itself, and the Sun in turn is held in orbit around the centre of the Milky Way by the gravitational grip of all the matter in the Milky Way galaxy. Secondly, as these examples highlight, there is only one kind of gravity, which adds up the more matter you have. Electricity, like magnetism, comes in two varieties – positive and negative charge, north and south poles – which cancel each other out. So, for example, although there is a lot of negative charge in your body (in the form of electrons) and a lot of positive charge (in the form of protons), overall your body has no electric charge and is neither attracted nor repelled by the person next to you, electrically speaking.* The other two forces, the strong and weak forces of particle interactions, only have very short ranges, and do not directly affect anything on scales larger than an atomic nucleus.

In order to see how weak gravity really is, we have to compare like with like, and look at how strongly the particles inside an atomic nucleus influence one another through the four different forces. If the strength of the strong force is defined as 1, then in the same units the strength of electromagnetism is $1/137$, which is about 10^{-2}, and the strength of the weak force is 10^{-13}, or one tenth of a thousandth of a billionth of the strength of the strong force. But the strength of gravity is only 10^{-38}. Even the weak force, which is by far the weakest of the other forces, is 10^{25}, or ten million billion billion, times stronger than gravity. The reason may lie in the way strings are connected, or not connected, to branes.

Once again, the idea came from a mathematician playing with the equations just for fun. The mathematician was Joe Polchinski, of the University of California, Santa Barbara; in the mid 1990s he was interested in what happens to open-ended strings. Although the original rubber band idea envisaged loops of string, string theory also allows the existence of strings with ends, and it wasn't obvious at first sight what happened to the ends of the strings. Polchinski had the idea that the ends of strings might be attached to a surface, but free to slide about over the surface. The surface would be a membrane,† and if the membrane had three dimensions the mathematics of vibrating strings attached to surfaces in this way could be used to describe the properties of 'particles' – in particular, the particles that carry the forces of nature. What mathematicians refer to as sliding about 'on'

---

* You can sometimes build up a small electric charge on your body, for example when walking about on a nylon carpet in some kinds of shoes on a dry day. The excess electricity gets discharged when you touch something metal, such as a door handle, giving you a noticeable shock.
† Polchinski calls them D-branes, in honour of the nineteenth-century mathematician Peter Dirichlet, who studied the way waves bounce off surfaces.

a three-dimensional membrane is, of course, the same as moving about 'in' the three dimensions of space that we perceive.

At first, this was no more than a curiosity. But after Witten came up with M-theory, it was quickly appreciated that our visible Universe might be a 3-brane floating in higher dimensions, with particles such as photons and electrons the detectable manifestations of strings attached to the brane and sliding about on (or in) it. The mathematics works as a description of the way particles interact – in 1994, for example, Polchinski showed that D-branes are possible sources of electric and magnetic fields. But there is one exception. Gravitons only exist as loops of string, and cannot attach to the brane. The only way that they can interact with the particles that are attached to the brane is by moving from one object out into the extra dimension(s) of space, and then moving back into the 3-brane to meet up with another object. But some of the gravitons might not return. Gravity would in effect leak away from the brane, and the strength of the force of gravity would be reduced as a result.

If the extra dimension(s) into which gravitons could fly were infinitely large, gravity might leak away almost entirely, so that nothing in the Universe would be held together. But if the appropriate extra dimension was compactified in the simplest way, just by being shrunk down so that there was only a limited amount of space available in the direction at right angles to all three of our familiar space dimensions, the gravitons could not go very far. If the extra dimension was made small enough, it would be hard to tell whether gravitons had actually left the 3-brane at all, and the strength of gravity would be just the right size to explain its relationship to the other forces of nature in our world. Why should it be compactified in this particular way? Because there are many ways to compactify space, as I shall discuss shortly, most of which would produce a force of gravity either too weak or too strong to allow the existence of stars, planets and people. As discussed in Chapter Two, we can only exist to notice such things in a Universe with the right balance of forces.

There's a bonus. If gravitons can leak out of our Universe, they can also leak in – so could other interesting particles, which might be detected at the Large Hadron Collider, but that's another story. Many 3-dimensional brane worlds can lie side by side in higher dimensional space, like a stack of sheets of paper on my desk. The universe next door might be microscopically close to us in the eleventh dimension, close enough for some of its gravitons to leak into our Universe and affect the way things move. The effect would be just as if our Universe contained unseen dark matter tugging on the visible stars and galaxies. And dark matter is one of the essential ingredients of the standard model of cosmology. *The* essential ingredient is the Big Bang – and membranes can account for that, as well.

WHEN WORLDS COLLIDE

Branes float about in the higher dimensions of space, just as a two-dimensional membrane, or sheet, can move about in three dimensions. Rather than being stacked up like sheets in a pile of paper, branes (whole universes!) behave in the higher dimensions like material objects in three dimensions – they can move about and collide with one another, or orbit around each other like the Moon orbiting the Earth, or planets orbiting the Sun. A better analogy than a stack of paper sitting quietly on my desk would be the turmoil of the same sheets of paper blown about in a strong wind. Although we cannot detect the extra dimensions of space directly, there are fields which ought to be affected by the geometry of higher-dimensional space and by the proximity of the nearest brane. These might one day be detectable; meanwhile, physicists have tried to find ways in which these fields might drive inflation.

Pushing two branes together takes energy, just as it takes energy to push two positively charged atomic nuclei together. But if branes did smash together at high speed, their energy of motion might be converted into other forms, just as nuclear energy is released when a heavy atomic nucleus is struck by a fast-moving particle and fissions. At the end of the 1990s, Georgi Dvali, of New York University, and Henry Tye, of Cornell University, suggested that in a head-on collision between branes, part of the kinetic energy of the collision might be converted into the kind of energy needed to trigger inflation. It turned out that the energy involved is far too small to do the trick; but that wasn't the end of the story.

A breakthrough came in 2001, when a large team of researchers started thinking about another way to get energy out of brane collisions. Just as atomic energy pales into insignificance compared with the energy that is released when matter particles and antimatter particles annihilate – in such collisions, *all* of the mass is converted into energy, in line with Einstein's famous equation – so, they reasoned, vastly more energy than that in an ordinary brane–brane collision would be released if a brane met up with an antibrane and annihilated. The equations allow for the existence of both branes and antibranes, and just as an electron is attracted to its antimatter counterpart, the positron, so a brane is attracted to an antibrane, ensuring that they will indeed collide and annihilate if they get close to one another.

This might not seem to be much use, since there would be no matter left over to inflate. But so much energy is released in brane–antibrane annihilation that some of it spills over into any nearby branes, providing more than enough spare energy to trigger inflation. There's a bonus. This process of annihilation naturally tends to produce a variety of universes with relatively few dimensions, like our own Universe. For example, if a 7-brane and its counterpart annihilate, they do not actually get completely converted into energy in one step. A lot of energy is released, together with fragments in the form of 5-branes and their antibranes. These fragments annihilate in their turn, leaving traces in the form of 3-branes and their antibranes. The 3-branes annihilate to make 1-branes, and only when the 1-branes annihilate is everything converted into energy. Large branes (that is, branes with many dimensions) fill up multi-dimensional space and quickly bump into one another. But small branes, like our Universe, are scattered more sparsely across higher-dimensional space, and can hang around for a long time before they are destroyed. This is a plausible reason why the kind of three-dimensional Universe we live in should be common.

On the other hand, there is something unsatisfactory about this idea, because it seems to imply a one-off evolution of the Multiverse from an earlier state with a small number of high-dimension universes to a later state with a large number of small-dimension universes. This is almost as uncomfortable as the idea of a unique Big Bang that just happened to produce conditions suitable for life. It would be much more satisfactory if we could take time out of the equation. One way to do this has echoes of the eternal inflation in an infinite de Sitter universe described in the previous chapter – but without inflation.

BY ITS BOOTSTRAPS

The leading advocates of this idea are Paul Steinhardt, of Princeton University, and Neil Turok, then of Cambridge University but now working at the Perimeter Institute, in Ontario. In 1999, neither of them were aware of Dvali and Tye's early work on brane collisions when they both attended a meeting in Cambridge where Burt Ovrut, from the University of Pennsylvania, gave a lecture outlining the idea of three-dimensional brane worlds separated from one another by minuscule amounts in a direction at right angles to all three of the familiar spatial dimensions – along the eleventh dimension, if you accept M-theory. Steinhardt and Turok were sitting on opposite sides of the room, but after the talk they converged on Ovrut, each with the same thought. Neither can now recall who came out with it first, but both clearly remember one of them blurting out, 'Can't these worlds move along the extra dimension? And, if so, is it possible that the Big Bang is nothing more than a collision between these two worlds?' It was the beginning of a long and fruitful partnership, described in their book *Endless Universe*.

This is still very much a work in progress; but the progress so far has produced an intriguing variation on the idea of a cyclic universe. Because the model is cyclic, we can start describing it anywhere in the loop, so I might as well start with the event that triggered the Big Bang. According to Steinhardt and Turok, this happened when two very smooth, flat and empty 3-branes came together and collided. Each brane would be a complete three-dimensional Universe like our own, with six additional compactified dimensions and the dimension of time, initially separated from one another by a tiny distance in the eleventh dimension. At first, they moved very slowly along the eleventh dimension, but as they got closer together they speeded up, pulled together by a spring-like force, until they collided with enough impact to heat both worlds to extreme temperatures. Crucially, though, the model does not require an input of energy big enough to trigger inflation. Instead, the temperature is 'only' about 10^{20} K, which is hot enough to explain how the particles of our world were made out of radiant energy, and how the hot fireball then cooled to produce atoms and the background radiation. Far from being infinite, the energy involved is much less than the energy associated with events at the Planck scale, so there are no singularities involved in this model of the early Universe.

Steinhardt and Turok also have a neat explanation of how the irregularities that grew to become galaxies got started. Just as in the inflationary model, it depends on quantum fluctuations. Because of quantum fluctuations, no spacetime can be completely flat and empty, and on the quantum scale the two colliding branes are inevitably wrinkled in a random way. As a result, the two branes do not come together everywhere precisely simultaneously; bits that stick out a little in the eleventh dimension will collide first, and these regions will get hot first. After the two branes have collided, they bounce apart, but in each brane, although the temperature is high everywhere, it is a little bit higher in some places than in others. This whole process of collision and bounce may take billions of years, rather than the split-second it takes for inflation to make a universe out of flat spacetime, but because in both cases the ultimate cause of the irregularities is quantum fluctuations, the pattern of irregularities you end up with in the colliding brane model is exactly the same as the pattern produced by inflation – which means that it is exactly the same as the pattern observed by the microwave detectors on satellites such as WMAP.* The pattern would, though, be the same in each universe, because the hot spots were where they touched each other first. So clumps of matter in one universe would match up with clumps of matter in the universe next door. As the branes moved apart, because of the spring-like force they would never get more than a microscopic distance from each other, close enough for each to exert a strong gravitational influence on the other as free-flying gravitons bridged the gap. On this picture, dark matter is matter in the universe

* There are refinements of this idea where the branes never quite touch, but repel each other when they are very close together. But the end result is the same.

next door, and next door is closer to us than the distance across an atom.

So colliding and bouncing branes can produce a universe indistinguishable from our own, emerging from a Big Bang. The spring-like force that holds the two branes together is related to dark energy, so, unlike inflation theory, the bouncing model* requires (rather than just permits) the presence of dark energy in our Universe. Just as in the dark energy models described earlier, we are living at an interesting stage in the life of the Universe, when dark energy is starting to dominate the expansion. As time passes, the Universe will expand faster and faster, with matter spread thinner and thinner, until spacetime is completely flat (except for quantum fluctuations) and almost completely empty, without even a single electron in a volume of space equivalent to the size of the entire observable Universe today. It will have become an expanding de Sitter space.

Meanwhile, the other brane world, having bounced away from our world in the collision, is undergoing its own expansion and thinning in the same way. But although the two worlds move apart along the eleventh dimension, they are still held together by the springy force, and eventually start to move towards one another again. This takes a very long time – Steinhardt and Turok talk of 'trillions' of years, where a trillion is a million million, or 10^{12}. That is, though, 'only' about a hundred times longer than the time that has passed so far since the Big Bang, which shows the power of runaway exponential expansion once it gets started. At last, the branes come together, and the whole process repeats. Phoenix-like, a new universe (or two) is born out of the ashes of the old. The Universe seems to be (re-)creating itself, lifting itself into existence by pulling on its own bootstraps, offering an endless cycle of universes in which the constants of nature may differ from cycle to cycle, so that we only live in a particular bubble where the conditions are just right for life. But where does the energy to drive each Big Bang come from? And how is all of this compatible with our ideas about entropy? Once again, it's all down to gravity.

---

* Steinhardt and Turok call it the 'ekpyrotic' model, from the Greek word for conflagration. I hate this ugly term, and shall try to avoid using it.

THE BOTTOMLESS PIT

Each time the two branes come together, some of their kinetic energy of motion along the eleventh dimension is converted into radiation and matter. This energy comes ultimately from the gravitational force between the two branes, helping the springy force pulling them together. Our everyday experience tells us that in that case the strength of the bounce between the two branes should wind down over many cycles, just as a ball dropped from a height onto a hard surface bounces a little less each time it hits the ground. It seems that the maximum separation between the two membranes should get less with each cycle in the same way. But this doesn't take account of the negativity of gravity.

Gravity isn't just negative – it is a bottomless pit. There is no limit to the amount of negativity, so there is no limit below which the energy cannot fall. This is quite different from, say, temperature. There *is* a limit to how cold something can be. Because we choose to set our zero of temperature at the freezing point of water, this lowest temperature comes out as a negative number, $-273.15°C$. But we can choose any point on the scale as our zero, and on the Kelvin scale this ultimate lowest temperature is set as zero (0 K), so all temperatures are positive and ice melts at $+273.15$ K. You cannot do this kind of thing with gravity, because there is no ultimate lowest energy point of gravity to measure from. Because of this, it isn't even possible to tell that there is less energy in each bounce than in the preceding bounce. All that could be detected, even in principle, would be the temperature, matter density and expansion rate at the closest approach of the two branes. Steinhardt and Turok assure us that their calculations say that these properties repeat precisely from one bounce to the next. The process is exactly cyclic.

Their model also solves the problems with entropy that occur in bouncing models that involve a reversal of the expansion of the universe. In the Phoenix model, in three-dimensional space the expansion never reverses, so these problems do not arise. There is no 'big crunch' in which matter is concentrated and the entropy built up earlier in the cycle prevents the bounce being a mirror image of the previous big bang. In the Phoenix model, entropy increases in the usual way as the universe expands away from each big bang, but the extreme stretching of space associated with the dark energy produces plenty of extra room for the entropy, keeping its density low. Even when the two branes are coming together along the eleventh dimension, equivalent to the contraction phase of the old cyclic models, three-dimensional space is still expanding. It never stops expanding. By the time the collision occurs and more matter and radiant energy are poured into the universe, the entropy density is vanishingly small, and the whole process can repeat. Infinitely negative gravity and infinitely expanding space provide the background for an infinite succession of big bangs. Instead of offering an infinite variety of universes across space, the model offers us an infinite variety of universes – anything that is allowed to come out of a big bang – in time. Steinhardt and Turok emphasize that as far as space is concerned, in each cycle it's a case of *plus ça change, plus c'est la même chose*.

THERE'S LOTS OF PLACES LIKE HOME

Although the Phoenix model produces a universe that is in many ways indistinguishable from a universe produced by inflation, there are some key differences between the two ideas, one of which may make it possible to tell which kind of universe we live in. Conceptually, the ideas are almost the opposite of one another. The inflationary idea suggests that a universe like our own, even though it looks infinite from the inside, is actually a rare bubble separated from similar universal bubbles by huge expanses of inflating space. There may be infinitely many of these bubbles, and they may provide every possible variation on the theme of what can happen as a result of inflation, but they are in no sense 'nearby', and there is no reason to think that any one of the bubbles chosen at random will resemble our bubble. Most of the Multiverse is very different from our home.

The Phoenix model, on the other hand, says that in each cycle everywhere is more or less the same – the local universe is typical of the universe as a whole. The whole of the infinite universe contains stars and galaxies distributed in the same general way as the stars and galaxies in our cosmic neighbourhood, and creatures living on any planet elsewhere in the infinite universe will look out into space and see the same sort of things that we observe. Everywhere is like home, even if not quite like Kansas.

This is not something that we can ever hope to test by observations, since in either case the maximum region of space that we could ever see even with perfect telescopes would look much the same as the region of space we have already observed. But there is one way in which the two models make different, detectable predictions. According to inflation, the extreme conditions that existed at the birth of the Universe should have produced intense gravitational radiation, ripples in the fabric of space, that has left an imprint on the cosmic microwave background radiation. The effect is calculated to be tiny, in terms of the observable effects on the radiation today, but it might just be detectable by instruments carried on board the European satellite Planck, launched in 2009 and now analysing the background radiation. On the other hand, the Phoenix model, being much gentler and less extreme than inflation, unambiguously says that there should be no imprint of gravitational waves in the background radiation.

If Planck fails to find the predicted effect, that will not prove the Phoenix model right, because the waves might still be there at a lower level. But if Planck does find traces of gravitational waves, that will definitely prove the Phoenix idea wrong.

Steinhardt and Turok like the simplicity and straightforwardness of their model:

The underlying mechanism driving the cycles is gentle and self-regulating. The collisions between the two branes occur at moderate speeds (well below the speed of light). The dark energy density is always low, and there is no runaway to high-energy states where vast expanses of inflationary universe are created. Instead, dark energy acts as a shock absorber that keeps the cycles under control, suppressing the effects of random fluctuations so that the regular, periodic evolution is kept on track.

They make a good case, but I'm not convinced; I expect that Planck, or one of its successors, will find traces of gravitational waves at work in the early Universe. The Phoenix model is attractive, but it has one big flaw. Why stop at two branes? What else is going on in 11-dimensional space while these two branes are locked in an eternal embrace? The most exciting thing about M-theory, and the most compelling reason to take the idea of the Multiverse seriously, is that it offers an infinite choice of possible worlds, not just one boring pair of cymbals repeatedly clashing out the same old song. Leonard Susskind has dubbed the variety offered by M-theory 'the cosmic landscape', and it is currently the hottest cosmological game in town. There may even be room in the landscape for cyclic universes, as one option among many rather than as a unique solution to the puzzle of our existence.

EXPLORING THE COSMIC LANDSCAPE

The reason why there is so much choice is that although it is true that there are only five different possible kinds of string theory, all contained within a single M-theory, there is a mind-bogglingly large number of different ways in which space can be compactified within each of those five theories. And each compactification corresponds to a different universe. It is the details of the compactification which determine things like the strengths of the forces and size of the constants of nature discussed in Chapter Two, the charge on the electron, and even things like the number of electron-like particles that can exist in any particular universe. In our world, there is just one kind of electron and three kinds of quark; but with a different compactification, there might be a world with, say, three kinds of electron and five quarks.

The reason for this variety of choice – which Max Tegmark classifies as a 'Type IV' Multiverse – is easy to see when you remember that there are six spatial dimensions to compactify (indeed, in some universes, more or less than six of the ten spatial dimensions might be compactified), and each can curl up in a variety of ways. The simplest example is when a dimension curls up into the equivalent of a tiny sphere. But it is also possible for space to curl up into the equivalent of a ring donut – a torus – or a figure 8 configuration like two donuts stuck together, and so on. The situation is more complicated with six dimensions to play with, and very hard to visualize, but all the possibilities can be described mathematically; they are known as Calabi–Yau manifolds. The holes in the manifolds, equivalent to the holes in lot of ring donuts stuck together side by side, are important, because fields can be threaded through the holes, wrapping around the ring part of the torus, or its equivalent in higher-dimensional space.

We are all familiar with the idea of magnetic 'lines of force', which show up when iron filings are sprinkled on a sheet of paper placed over a bar magnet. Such lines of force, for other fields as well as magnetism, can wrap round the higher-dimensional 'donuts' and through the holes. But there is a limit to how often a field can wrap round in this way, because the field lines inside the holes produce an outward pressure, and if this is big enough it will make the holes expand, de-compactifying the space. The interaction of the fields and the curled up dimensions produces an energy, which is the vacuum energy associated with a particular configuration when no matter is present – the energy of empty space. This is the key to understanding our place in the cosmic landscape.

When physicists first began to consider the implications of all this for cosmology, in the late 1990s, they were confronted with the puzzle of why our particular Universe should have crystallized out from the many possibilities allowed by this deeper understanding of M-theory. Some people were tempted by the idea that a form of anthropic principle might be at work, if there were many different possible configurations all existing in the Multiverse; but that would require a huge number of possible configurations to make it at all likely that a universe like ours would exist, and at first nobody knew if this was allowed by the theory. In the year 2000, Raphael Bousso, then working at Stanford University, and Joe Polchinski put this idea on a more secure footing by pointing out that quantum fluctuations could allow regions of space to jump from one compactification con-figuration to another, on very rare occasions. This would, in effect, create new bubble universes, each emerging from its own big bang and expanding in its own way and with its own set of physical laws, separated by regions of expanding space with the 'old' configuration, making a pattern similar to that of the bubble universes in eternal inflation. There could be bubbles within bubbles within bubbles, with no beginning and no end.

This didn't catch on immediately, but in 2003 a team known as KKLT* came up with a realistic estimate of the total number of universes allowed by compactification of some of the dimensions of ten-dimensional space. They found that, in round numbers, the compactified space can contain 'donuts' with up to 500 holes, but not many more. They also discovered that each hole can only be threaded by a few lines of force, certainly no more than nine. Choosing nine as the maximum number gives ten possibilities for each hole (0, 1, 2 . . . 8, 9), and with 500 holes that means that the maximum number of configurations for the vacuum is 10^{500}. For comparison, there are only 10^{80} atoms in the entire visible Universe. This huge number, 10^{500}, is not infinite, but it is amply big enough to make it entirely likely that if all these configurations exist in the Multiverse, some of them will be very like our Universe.†

Leonard Susskind became an enthusiastic supporter of the idea, and gave this Multiverse of different universes with different compact configurations and different energies the name the Cosmic Landscape. It's under that name that the idea caught on in the years after 2003. In an attempt to put the mind-boggling number 10^{500} in perspective, Susskind points out that if you placed a series of dots one Planck length apart across the diameter of the entire visible Universe, there would only be 10^{60} points in the line. Then he gives up. 'The number 10^{500} is so staggeringly large that I can't think of any way of graphically representing so many points.' But how about this. Volume goes as the cube of radius, so even if you filled the visible Universe with points in this way, there would only be about 10^{180} of them. You would need 10^{320} entire universes just like our own visible Universe to contain 10^{500} 'Planck points'. That's the amount of variety in the Cosmic Landscape. This extraordinary number is still not the end of the story. It doesn't mean that there are 'only' 10^{500} different universes in the

* From their initials – Shamir Kachru, Renata Kallosh, Andrei Linde, Sandip Trivedi.
† Some calculations suggest that there could be as many as 10^{1000} possible vacuum states; 10^{500} is actually a conservative estimate!

Multiverse, but that there are 10^{500} different *kinds* of universe. There is still scope to have many copies of the same kind of Calabi–Yau manifold, with the same physical laws, each developing in its own way from its own 'big bang'; the implications of this will be discussed later.

Even so, the Cosmic Landscape is not equivalent to a smooth continuous surface. It would take an infinite number of points to fill up such a surface entirely. The image you should have is more like a pointillist painting. From a distance, the landscape looks like one of smoothly rolling hills and valleys, perhaps with occasional high peaks and deep chasms, but close up you can see that it is made up of a myriad of tiny dots almost touching one another – not unlike, in fact, the way a seemingly continuous smooth sheet of paper is made up of a finite number of atoms.

But not all points in the landscape are equal. Crucially, they do not all have the same energy. The idea of a landscape is one frequently used in science as a way of visualizing some kind of variable property, commonly energy. Susskind says that he knew of the idea from its use in chemistry in connection with the energy of large molecules. A molecule that contains thousands of different atoms can in principle be put together in many different ways, with the same atoms arranged in different patterns. Each variation on the theme has its own energy, and the range of possibilities can be represented as that rolling landscape with hills, corresponding to higher energy configurations, and valleys, corresponding to lower energy configurations. This enables chemists to get an idea of how stable a particular configuration is – if it sits on top of a hill, then any disturbance is likely to make the atoms re-arrange themselves into a configuration with lower energy, in effect rolling down into a valley. But if the configuration is already at the bottom of a valley, it will be stable, and unlikely to change its shape.

In the Cosmic Landscape, the hills and valleys correspond to the vacuum energy for each compactification configuration. The idea is that when a quantum fluctuation produces a 'new' configuration it does so at random. If the new space is in a valley, all well and good. It stays there, with a particular set of physical laws, and does its own thing. But if the new configuration is on top of a hill, or high up on the side of a valley, it will roll down towards the lowest point, giving up energy as it does so, before settling into a stable state. This rolling, with a release of energy, is exactly the situation that we already know as inflation.* The energy drives a big bang, and the low level of vacuum energy at the bottom of the valley is the dark energy (the cosmological constant) that eventually makes the expansion of that particular universe accelerate, providing immense amounts of new space in which rare quantum fluctuations can make new universes. 'All of known cosmology,' says Susskind, 'took place during a roll from one value of the cosmological constant to a much smaller one.'

If you imagine a table-top model of such a landscape, with hills and valleys, then roll a marble on to the surface, the marble will end up in a valley, not on a hill top. Stable universes like our own must have small values of the cosmological constant. But why should our particular Universe be picked out? It isn't very likely, if you roll just one marble on to the landscape. But if you roll very many marbles in at the same time, or if you keep rolling marbles in one after another for eternity, then all of the valleys will be occupied. Each point on the string landscape corresponds to a particular compactification with its own set of physical laws. But according to Bousso and Polchinski, there are about 10^{380} vacuum states in 'the sweet spots' where universes more or less like ours can exist. Although we can only exist in such a universe, that's ample to satisfy the requirements of anthropic cosmology. String theory provides a natural reason for the Multiverse to exist, and eternal inflation provides a natural process to occupy every possible valley in the Multiverse landscape. The laws of physics depend on where in the cosmic landscape a universe is located, and on this picture, cosmologists like to quip, the laws of physics that we know may just be local by-laws.

---

* Because of the 'pointillist' nature of the landscape, this 'rolling' actually involves a kind of quantum jittering, like a ball bouncing down a staircase, but that doesn't affect the argument.

THE RETURN OF SCHRÖDINGER'S CAT

This isn't quite the end of the story (so far) of the Cosmic Landscape. The number of different kinds of universe that populate the string landscape is so enormous that the distinctions between them are not always very great. There is room in the landscape, for example, for a universe where the laws are exactly the same as in our Universe, but the electron has a tiny bit more (or a tiny bit less) mass. There are also, of course, many more universes which differ wildly from our own. But cosmologist Paul Davies, now based at Arizona State University, says that 'It is not too much of an exaggeration to say that you could dream up a universe, choosing whatever sort of low-energy physics takes your fancy (within reason), and there will be a universe somewhere matching that description among the unimaginably vast smorgasbord of possibilities.' By 'within reason' he simply means that you cannot, for example, have an inverse square law of gravity in a universe with any number of large dimensions, only in three-dimensional spaces, as discussed in Chapter Two. There are some constraints on what compactification can produce in the way of different universes, which is why there are only 10^{500} points on the cosmic landscape, not an infinite number.

There is also room in the cosmic landscape for universes in which the laws of physics are indistinguishable from those in our Universe, but the histories are different. Where, for example, the Earth was not struck by a large object from space some 65 million years ago, and the era of the dinosaurs was not brought to an abrupt end. Sound familiar? This is very like the Everett Many Worlds Interpretation of quantum mechanics. So much like it, indeed, that Susskind has suggested that it is essentially the same thing.

Susskind points out the similarities between the two ways of looking at the Multiverse of string theory (which he refers to as the 'megaverse') and the two most familiar ways of looking at the interpretation of quantum theory. One point of view is that of a single observer, who we imagine to be impervious to the changes in the structure of space, watching things from within a single causal patch. From this point of view, there is a single 'pocket universe' which has a certain set of physical properties. In time, quantum fluctuations akin to tunnelling convert the space into a different form, with a different set of physical laws, and most probably with lower energy. This process happens over and over again, but the vast majority of the succession of universes the observer passes through are inhospitable to life and sterile. It's like a single river meandering across the Cosmic Landscape, only passing through a narrow region of the Landscape and leaving the rest unexplored. The chance of the hypothetical observer ever seeing a universe suitable for life is very small. Susskind calls this the 'serial' view, since successive universes follow one after the other.

This is like the Copenhagen Interpretation of quantum physics, the view developed chiefly by Niels Bohr, in which whenever there is a choice at the quantum level one path is chosen and all the other possibilities disappear in the 'collapse of the wave function'. But as Susskind emphasizes, the collapse of the wave function is not part of the mathematics of quantum physics, it is 'something that Bohr had to tack on in order to end the experiment with an observation'. Since the universe does not end with an observation, the Copenhagen Interpretation, quite apart from its other flaws, is inappropriate when we are trying to describe the Multiverse.

The other approach is what Susskind calls the parallel perspective. Instead of a single universe passing through different states allowed by string theory one after the other, this involves many pocket universes spread out across the Cosmic Landscape, developing in their own ways at the same time – in parallel with one another. From this perspective, it is absolutely certain that some will end up in states suitable for life. 'Who cares about all the others?' asks Susskind rhetorically. 'Life will form where it can – and only where it can.'

This, he says, is like the Everett version of quantum physics. It's even more like David Deutsch's variation on that theme. Starting from any particular pocket universe in the Cosmic Landscape, every possible future will exist somewhere in the Landscape, and there is no unique thread linking a sequence of pocket universes meandering like a single river across one narrow region of the landscape. It's more as if a deluge flooded the entire Landscape so that every hollow became filled with water.

Susskind sees this as a profound insight:

Perhaps in the end we will find that quantum mechanics makes sense only in the context of a branching megaverse and that the megaverse makes sense only as the branching reality of Everett's interpretation.

Whether we use the language of the megaverse or the many-worlds interpretation, the parallel view, together with the enormous Landscape of String Theory, provides us with the two elements that can change the Anthropic Principle from a silly tautology into a powerful organizing principle.

This pulls together everything discussed in this book so far in such a pleasing way that it is tempting to end it here. The Cosmic Landscape of string theory is just the many worlds theory of David Deutsch writ large, and with inflation included within itself. But there is one loose end that needs tying up, and one more piece to add to the picture put together by Susskind. The loose end, taken seriously by some people, is the idea that the Universe is a fake – that we are living in a computer simulation, not unlike the world of the characters in the *Matrix* series of movies. The last piece of the puzzle is the role played by black holes in the birth of universes.

7

Faking It? Or Making It?

In the early twenty-first century, two startling ideas have emerged for serious consideration as an explanation of the Universe we see around us, within the context of the Multiverse of string theory. The first is that our universe is a computer simulation – in other words, a fake. This idea is taken seriously by a surprisingly large number of eminent cosmologists, but in my view they are barking up the wrong tree, for reasons I will explain. The second idea is less fashionable, but in my view far more compelling. It is that our Universe is an artificial construct, manufactured deliberately by intelligent beings in another universe. There is a huge difference between faking it and making it, and the evidence suggests to me that making it is much more likely. These may seem at first sight more like philosophical or religious matters than real science. But the science they are based on, the science of the string theory Landscape, is real enough. Because the conclusions are so important, it's worth running through this to make the scientific basis of the argument clear.

IS IT SCIENCE?

There is a widespread impression that the Multiverse concept is some kind of vague philosophical idea and that the anthropic solution to the puzzle of cosmic coincidences, such as the value of the vacuum energy that drives the acceleration of the Universe, is some kind of tautology, or circular argument. Raphael Bousso explains why this is not the case particularly neatly, and I have borrowed the following three-step argument from him.

First, either there is a Multiverse, or there isn't. That is clearly a scientific question, not a matter of opinion or what you would like to believe. The crucial point in the scientific argument is whether the rules of particle physics allow for the existence of at least one 'false vacuum' – that is, a state of empty space which has more energy than another state of empty space. If it does, then it is certain that sooner or later a quantum fluctuation will produce such a false vacuum. The false vacuum will then 'roll down' to the lower energy level, driving inflation and making an infinite expanding space within which quantum fluctuations will produce more expanding bubbles, each with infinite volume. As long as there is a single false vacuum state allowed by the laws of physics, an infinite Multiverse follows, even if you start with a finite volume of space. 'The idea that the [metaverse] may be infinitely large, and that it may contain infinitely many copies of our own region, is,' says Bousso, 'something that arises reliably from a simple and reasonable assumption.'

Secondly, the modern version of the Multiverse idea requires a Cosmic Landscape that contains not one but very many false vacua, making a complicated Multiverse with many different kinds of universes. This is a great help in explaining such things as the size of the cosmological constant or the values of the carbon and oxygen resonances using anthropic arguments. But it isn't obvious how to make such a Landscape from first principles. Yet string theory produces just such a Landscape, involving D-branes and multi-dimensional donuts looped with fields. 'It's truly remarkable, in my view, that string theory, which gives you essentially no freedom in choosing parameters, has turned out to provide just the ingredients that allow for a realization of [this kind of Landscape]. This didn't have to happen, but it did.'

Finally, because the Landscape is not simply a metaphysical notion, but a concrete model arising from a specific theory of physics, there is no reason why, in principle, it should not be possible to make testable predictions from it. This is not going to be easy, and even the technology of the Large Hadron Collider may not be up to the task of testing such predictions, but as Bousso emphasizes it is not the case that 'anything goes'. He makes an analogy with the behaviour of atoms and nuclei. For example, the relevant theories of physics do not allow there to be more than about a hundred protons in an atomic nucleus, and iron cannot be gaseous at room temperature. He hopes, and expects, that there will be ways to extract such broad rules of the behaviour of matter at what are low energies compared to the Big Bang, but high by the standards of everyday life, from string theory. This would be like, for example, the way the electrical conductivity of iron depends on the bulk properties of many atoms, without it being necessary to calculate from first principles the interactions between every proton and every electron in a lump of iron.

String theory is real science, and the Cosmic Landscape is part of string theory. So where, literally, does that leave us?

INSIDE INFORMATION

There's nothing new about making a kind of image of the world that we can view from the outside – even without using live actors, animated cartoons have been doing it since the days of 'Steamboat Willie' and his contemporaries. These simulations have progressed from hand-drawn pen-and-ink sketches to computer simulations that are very close to looking like the real world. There is every likelihood that before too long, possibly by using holographic techniques, these simulations will even look three-dimensional, so that the viewers can be immersed in the scenery or the action, and be fooled into thinking they are experiencing the real thing. But there is an enormous difference between simulating the appearance of the world outside well enough to fool our senses (not that we have even got there yet!) and simulating the whole world and everything in it, including the thought processes of any intelligent beings 'living' in that simulation, inside a computer. Leaving aside the technical difficulties of how this might be achieved, the essential question that has to be answered is how much information, in the form of binary digits (bits) would be needed to simulate a universe like our own in this way – or, looking at it from the other side, how much information, in terms of binary digits, does our Universe contain?

Cosmologists are intrigued by the concept of information and its relationship to the Universe, partly because of the peculiar way that information (which is closely related to entropy) interacts with black holes, and partly because, as Jacob Bekenstein has summed it up, 'a final theory must be concerned not with fields, not even with spacetime, but rather with information exchange among physical processes.' On this picture, even M-theory would be some cruder representation of a flow of information, although that would not invalidate any of the conclusions drawn from M-theory. This is not the place to look in detail at the information debate, but what matters here is that this interest in information and information flows has provided a lot of relevant input to the discussion of how to fake a universe.

Bekenstein, who now works at the Hebrew University of Israel, made his name in the 1970s with his work on the relationship between the entropy of a black hole and its temperature. This was later developed by Stephen Hawking, of the University of Cambridge, who showed how it would lead to a constant stream of particles and radiation away from the surface of a black hole; this became known as 'Hawking radiation'. A large black hole swallows up more mass-energy than it radiates in this way, so it grows in spite of the radiation, but a tiny black hole would shrink as it lost energy, radiating faster and faster as it shrank away nearly to nothing before exploding in a final shower of particles.

All this poses many puzzles about information and the Universe. What happens to information that falls into a black hole? Suppose it is in the form of a complex artefact like a spaceship. Has the information about the spaceship been lost? And has there been a corresponding change in the entropy of the Universe, or do we have to take account of the entropy of the black hole, which now includes a component due to the spaceship, as well? For a long time, Hawking was a leading proponent of the idea that information that fell into a black hole is lost forever and cannot be retrieved. On the other side of a fierce but friendly debate, a group which included Leonard Susskind argued that there is a 'law of conservation of information' and that information could not be lost from the Universe even by disappearing inside a black hole.

Susskind and his colleagues reasoned that although the information might be scrambled up so that it was almost impossible to decipher it, the bits of information that fall into the black hole are returned to the outside Universe in the form of changes in the Hawking radiation. Susskind says that this is like shuffling a pack of cards. If you start with a pack of cards with all the suits in sequence and the cards in numerical order within the suits, or in some other arrangement, then shuffle it in some complicated way but following a definite rule, the information is still there, although it is impossible to retrieve the original pattern unless you know the rule for 'reverse shuffling' the pack to get back to its starting position. This is unlike the situation in which the pack is shuffled at random, because then there is no rule for doing the reverse shuffle. As long as the reverse-shuffle rule exists, even if you don't know what the rule is, the information is not lost.

The result was a classic stand-off between the general theory of relativity and quantum theory. The general theory says that information falls into the hole; quantum theory says that, in Susskind's graphic words, it is 'as though the message were torn from the hands of its messenger and transferred to the outgoing Hawking radiation just before passing the point of no return'. Like so many quantum dilemmas, the resolution of the dichotomy is to have your cake and eat it. Because of the way time gets distorted by gravity, to anyone in the outside world monitoring what goes on as an object falls into a black hole, what they see is the object getting closer and closer to the hole but moving more and more slowly, never crossing the point of no return known as the event horizon. Just outside the horizon, the object is eventually destroyed and turned into outgoing Hawking radiation. But anyone falling in to the hole, perhaps riding in the hypothetical spaceship, doesn't experience anything like this. If the hole is big enough, the spaceship and its passenger pass through the horizon and beyond the point of no return in complete comfort, without feeling a thing. The passenger can watch the world around the spaceship, even inside the black hole, until it reaches the central region where spaceship, passenger and all are torn to pieces by tidal forces.

Susskind and the Dutch physicist Gerard 't Hooft showed in the 1990s that both these views are valid, in the same way that both the view of light as a particle and the view of light as a wave are valid. This is a form of so-called complementarity, an idea entirely familiar to physicists from their studies of the quantum world, where entities such as electrons can be described either as waves or as particles, and both 'complementary' descriptions are correct. Even Hawking now agrees that he was wrong, and information is not lost when an object falls into a black hole. As Susskind puts it, 'to someone outside a black hole, the events in the life of the trans-horizon explorer are behind the horizon. But those events are physics, not metaphysics. They are telegraphed to the outside in scrambled holographic code in the form of Hawking radiation . . . it doesn't matter if the code [rule] is lost, or even whether we ever had it. The message is in the cards.' And there could be scrambled messages coming in this form from other universes if, as seems likely, black holes are passageways to other worlds. In the Multiverse, information can shift from one region of spacetime – one universe – into another through wormholes, so that in the entire Metaverse information is never lost.

One of the fruits of the investigation of the interaction between black holes and information/entropy has been a calculation, made by Bekenstein, of the total information content of a black hole. This is proportional to the surface area of the horizon, marking the point of no return, that surrounds the hole. Logically enough, if a small black hole contracts as it radiates energy and information away the surface area gets smaller, while as a large black hole swallows things packed with information it grows and the surface area gets bigger.

If the surface area of a black hole event horizon is divided up into squares each with sides as long as the Planck length, and each square contains one bit of information, the total number of squares is a measure of the information content of the black hole. A black hole a centimetre or so across would contain about 10^{66} bits of information. The big question is, how much information does the Universe contain, and how big a black hole would you need to store it in the most efficient way possible? Bekenstein estimates that the *minimum* amount of information in the Universe is equivalent to 10^{100} bits, a number known as a googol. The search engine Google uses a similar sounding name, chosen to make it sound impressive, but it cannot access anywhere near that much information. To store a googol bits, you would need a black hole one tenth of a light year across.

That's the minimum possible size of the memory store needed for a computer running a simulation of the Universe, regardless of the size of the computer itself, or whether or not it is operating on quantum principles. Compared with the size of the Universe, this is surprisingly small; but the calculation is open-ended at the top end, and there is no known upper limit to the amount of information the Universe contains. It is quite possible, and I believe much more likely, that the smallest thing that can store enough information to describe the universe is – the Universe itself. After all, from the perspective of bubble universes in the metaverse, the Universe *is* a black hole, and its surface area is then the minimum size corresponding to the amount of information it contains. This would make life very difficult for the fakers!

THE FAKERS

The argument used by proponents of the fake universe idea is essentially that in principle it is so easy to simulate a universe in a computer that there must be very many simulations scattered across the Multiverse, made by programmers in real universes and perhaps including simulations made by simulated programmers in their own simulated universes, simulation within simulation like nested Russian dolls. Therefore, they argue, simulations far outnumber real habitable universes, so it is much more likely, statistically speaking, that we are living in such a simulation than that we are living in a real, physical universe. Simulations might be programmed just for fun, like the computer games in the Sim World series, or out of scientific interest in how a universe would behave if the constants of nature were changed slightly, for entertainment, or for reasons unintelligible to the human intellect. All that matters is that they are made.

You may wonder why anyone would think that it is easy to simulate a universe like our own, when the very minimum requirement is the ability to access information on the Planck scale from the horizon of a black hole one tenth of a light year across. But it all depends on what you mean by 'easy'. Simulating a universe may require the emergence of a super-civilization with computational power far exceeding anything we have been able to achieve, and it is possible that no civilization with this capability will ever emerge in our Universe. But the Multiverse allows for every possible kind of universe, including universes very similar to ours in which, through some rare, freak combination of events just such a super-civilization arises. Unless there is actually a law of physics which prevents the emergence of such civilizations at all, not just in one Universe but in every possible universe, then inevitably there will be universes very similar to our own in which this kind of super-civilization exists. It is very hard – impossible, as far as I am concerned – to imagine any law that could forbid this. Proponents of the simulation idea argue that even if such universes are rare, the super-civilizations they contain will produce a huge number of simulated universes, so that fake universes become overwhelmingly more common than real universes. In which case, the odds are that we are living in a simulation.

David Deutsch, who knows more about computation than anyone else I have met, is a particularly scathing critic of this kind of reasoning, which he describes as a 'chimera' and an 'untestable conspiracy theory'. He says that 'computation cannot explain hardware,' and points out that in the real world (and he is convinced that it *is* real), steam engines are possible but perpetual motion machines are not. Yet 'the quantum theory of computation knows nothing of the second law of thermodynamics: if a physical process can be simulated by a universal quantum computer, then so can its time reverse,' and people 'living in' a simulation would be able to build perpetual motion machines.

John Barrow, a Cambridge theorist who has also thought deeply about this idea, takes a different tack. He argues that if we are living in a simulation, there ought to be clues in the form of 'glitches' in the laws of physics. Even a super-civilization, he reasons, may not know everything about the laws of physics. Indeed, one of the reasons for running the simulation might be to find out more about those laws. So they may include gaps, or errors, in their programming, which would show up in our world, producing puzzling experimental results – for example, observations might seem to show that the constants of physics are slowly changing as time passes, or are different in one part of the Universe than in another. Eventually, as he says in *The Infinite Book*:

[The simulations] would fall victim of the incompetence of their creators. Errors would accumulate. Prediction would break down. The world would become irrational.

It's an uncomfortable prospect, but I wouldn't lose any sleep over it. In the first place, there is no compelling reason to think that a Universe like ours can be simulated in any way except by making an exact same-size copy of our Universe. Apart from the 'bubble universe' problem already mentioned, there are real difficulties in storing the amount of information needed to specify the position of even a single particle in space. There are circumstances in which specifying position precisely involves an irrational number, such as π, which has an infinite number of digits after the decimal point. So you would need an infinite storage space to specify the number, and therefore the position of the particle, precisely!*

Of course, any computer programmer knows what you do in such a situation – you make an approximation. For some calculations you can use a value of $^{22}/_7$ for π, if you only need a rough answer. Or you can use 3.14159 to get a more accurate answer. The definition of π is the circumference of a circle divided by its diameter; it would be really interesting if some exquisite measurements of a perfectly drawn circle found that, millions of places after the decimal point, the number representing this ratio repeated in regular fashion, or came to an end. That would be powerful evidence that we live in a simulation.

But this idea of approximation raises another objection to the simulated universe idea. Why bother simulating the whole thing? Rather than simulating the complete workings of every distant quasar and star, why not simulate, say, the planet Earth in detail, with fake information coming into telescopes and other detectors on Earth in the form of photons that seem to hail from far away across the Universe? That would surely be ample if you were interested in something like the evolution of life on a single planet. Taken to its logical extreme, this argument takes us back to something very much like Boltzmann's brain puzzle. The fakers' argument rests on the assumption that what it is easy to make will be common in the Multiverse. The simplest simulation consistent with everything you have ever

* I discuss this problem more fully in *Deep Simplicity*.

experienced (including reading this book) is that your sentient personality is only a computer program running in a sophisticated machine, being fed data which give you the illusion that there is a world around you. Since that is easier than simulating an entire universe, or even a single planet, by the logic used by the fakers there should be many more such simulated 'brains' than there are complete simulated universes, so it is far more likely that you are merely an isolated but sophisticated computer program than that the Universe exists, either as a simulation or as a real world.

But even this isn't the end of the road. Just as it turned out to be easier to make entire universes rather than isolated Boltzmann brains, so, for different reasons, it is actually far easier to make entire new universes than it is to make simulated universes, or even simulated personalities. Nature does it all the time, through the medium of black holes. The propagation of universes in profusion needs no intelligent civilization to interfere; but if there is such intelligence, it could cultivate universes and propagate them the way a gardener cultivates and propagates plants.

BLACK HOLES AND BABY UNIVERSES

The physics of the events inside a black hole, collapsing towards a singularity, are exactly the time-reversal of the physics of the Big Bang, expanding outwards away from a singularity. In both cases, no physicist believes that there really is a singularity, but that something happens close to what we think of as the singularity to give the illusion that the collapse or expansion is proceeding all the way to or from a mathematical point. That something involves quantum gravity, events at the Planck scale where space itself has a foamy structure.

One of the most likely possibilities is that a spacetime collapsing towards a singularity will 'bounce' just short of the singularity itself, with the contraction turning into an expansion. This was the basis of early ideas about cyclic universes, which run into difficulties with entropy being carried over from one bounce to the next in the same spacetime. Those ideas involve a bounce occurring in our familiar three dimensions of space plus one of time, so that the collapse involves a shrinking of three spatial dimensions, with the same three space dimensions expanding again after the bounce. Quite apart from the puzzle of what happens to the arrow of time in such a situation, this cannot be what happens inside a black hole in our Universe, which is cut off from the three spatial dimensions outside the event horizon and can never return, except by evaporating away to nothing. But in the 1980s some mathematical physicists began to investigate the possibility that material inside a black hole that falls in towards the singularity might be diverted through a kind of spacewarp into another set of dimensions, that expand to become another spacetime.

This new spacetime is easy to describe mathematically, in terms of four dimensions, three of space plus one of time, which are *each* at right angles to *all* of the four dimensions of our own spacetime. With slightly more mathematical effort, the description can be extended to deal with compact dimensions as well. There is always the possibility that compactification may work out differently in the 'new' spacetime, but this is not something I want to dwell on here, since what matters for people like us is the existence of spacetimes like ours. From the perspective of Susskind's Cosmic Landscape, a black hole can be thought of, if this idea is correct, as a tunnel linking one part of the landscape with another. Every 'singularity', which means every black hole, is the entrance to another set of spacetime dimensions, and possibly another variation on the compactification theme, within the landscape of the Multiverse.

Another way to picture this is to start from the familiar analogy between our expanding three-dimensional space and the expanding two-dimensional surface of a balloon that is being steadily filled with air. Forget about the volume inside the balloon, and just think of the skin of the balloon as equivalent to space. On this picture, a black hole can be represented by a tiny blister on the surface of the balloon, which gets pinched off and starts to expand in its own right. There is a tiny 'throat' linking the original balloon with the expanding blister, and the blister can become as big as the original balloon, or bigger, without the structure of the original balloon (the original spacetime) being affected. All that inhabitants of the original universe can see is a black hole, but the black hole is one end of a kind of umbilical cord linking the parent universe to a new baby universe. In the new universe, to any intelligent observers the other end of the umbilical cord appears as their own big bang, complete with inflation and the production of profuse amounts of matter and energy thanks to the negativity of gravity.

There is, of course, no need to stop at one bubble. New bubbles of spacetime can form wherever there are black holes in the original universe, and new bubbles of spacetime can form in the same way anywhere in any of the new universes, the babies that the original space has given birth to. The most profound implication of this insight is that our Universe may have been born in the same way, from the collapse of a black hole towards a singularity in another part of the Cosmic Landscape. Our Universe has to be seen as just one component in a vast (presumably infinite) array of universes connected by tunnels through spacetime.*

Now, of course, anything possible can and will exist somewhere in the Multiverse, so in that sense there is no puzzle about the existence of such a network of interconnected universes, some of which are like our own. But it would be nice to know how the whole thing got started, and whether universes like the one we live in are sufficiently common that anthropic explanations for the cosmic coincidences are valid. The person who has thought most deeply about this is Lee Smolin, in his spare time from puzzling over quantum loop gravity. In the early 1990s he developed the idea, later elaborated on in his book *The Life of the Cosmos*, that a process of evolution, closely analogous to the way evolution works among life on Earth, may have produced a proliferation of universes like our own, starting from the smallest possible quantum fluctuations on the Planck scale. The two essential features of evolution are that individuals produce offspring which have characteristics that are similar to, but not identical to, those of their parents; and that the success of those offspring in reproducing in their turn depends on their own set of characteristics. So characteristics that aid successful reproduction tend to spread. Smolin says that this applies to universes as well as to life on Earth.

---

* And that infinite network is itself presumably just one of an infinite number of such networks to be found in the Cosmic Landscape.

SELECTING UNIVERSES NATURALLY

Smolin's key idea is that when a black hole in one spacetime collapses and tunnels into another spacetime, the physical properties of the 'new' spacetime, such as the strength of gravity and the values of the other constants of nature, are very nearly, but not quite, the same as in the parent universe.* This is like the way in which small mutations in DNA mean that the genome of a baby animal or plant is very nearly, but not quite, the same as the genome of the parent(s). Given such a situation, he argues that universes in which the constants of nature have values which encourage black holes to form will be more common than other universes, because they will have many progeny and those progeny will also have properties which encourage black holes to form. The same kind of arguments as those used in anthropic cosmology then imply that our Universe has been selected from all the possible universes in the Multiverse *not* because it is a good home for life, but *because it is good at making black holes*. The fact that what is good for black holes is also good for life forms like us is just a coincidence. In Smolin's words:

The parameters of the standard model of elementary particle physics have the values we find them to have because these make the production of black holes much more likely than most other choices.

But it is the slight variation in properties from one generation to the next which makes universes like our own overwhelmingly common.

Smolin explains how this happens using a version of the landscape idea. Where Susskind borrowed the idea from chemistry, Smolin borrows it from evolutionary biology – in fact, Smolin was using the idea in a cosmological context in the 1990s, some years before Susskind independently came up with his version. Smolin's landscape is imagined to be like a more or less flat plain, dotted with hills. A

---

* This idea of 'reprocessing' the parameters of physics actually goes back to the work of John Wheeler, in the 1970s, in the context of a single 'bouncing' universe. It was Wheeler, by the way, who gave black holes their evocative name.

point on the landscape corresponds to a particular set of values for the parameters which determine the properties of a universe, in particular its ability to make black holes. On the flat part of the plain there are universes with different detailed properties, but which have one thing in common – each of them makes only one black hole, emerging from the quantum foam and collapsing almost immediately. Small hills correspond to universes which live long enough, and have the right properties, to make a few black holes, and therefore a few offspring, while tall mountains correspond to regions of the Multiverse where universes have the right properties to make many black holes and have many offspring.

One important point, as with all variations on the cosmic landscape theme, is that there can be more than one universe at each point in the landscape. In the biological equivalent, a point on the landscape might correspond to the characteristics (or genome) of, say, a zebra; but there can be many zebra who all share those family characteristics. The landscape is such a familiar idea in biology that Richard Dawkins used it in the title of one of his best books, *Climbing Mount Improbable*. 'Mount Improbable' is a high peak in the evolutionary landscape, corresponding to the properties of a complex biological system such as the eye. A common objection to the idea of evolution by natural selection often made from a position of ignorance can be summed up as 'what use is half an eye?' In other words, how could a complex organ like the eye have evolved by a series of small steps, as evolution by natural selection requires. The quick answer is that half an eye can be very useful indeed, and certainly better than no eye at all when you are being threatened by a predator. The long answer, how eyes can evolve step by step starting from a patch of light-sensitive cells on the skin, is spelled out in *Climbing Mount Improbable*. This is a classic example of what the landscape is all about – you don't get to the top of a mountain in one leap, but by following a long and gentle path to the summit. If you see a picture of a person on top of a mountain, you don't assume they got there by magic, or a miraculous intervention by God, but that they climbed up step by step. So it is with eyes, and other complex organs. Eyes are so valuable, indeed, that they have evolved at least forty separate times, following forty different winding paths to this particular mountain top. If Smolin is correct, the complexity of the Universe around us emerged in a similar step-by-step process, starting from a simple quantum fluctuation.

In a simple example used by Smolin, we are asked to imagine that there are only two possibilities – that a universe can leave behind either just one offspring or exactly ten progeny. If we start with a single universe chosen at random from anywhere in the landscape, it is most likely that it will be on the plain, one of the kind which has only a single offspring. The offspring, like its parent, also pops into existence for a short time before collapsing and producing another single offspring, but crucially each generation has slightly different properties to the previous one. This is equivalent to the successive universes forming a path wandering across the plain. Eventually, a universe is produced with properties which leave it right next to a tiny hill corresponding to universes with ten progeny. One more tiny 'mutation' takes it up the hill, and the next universe in the chain does indeed have ten progeny. Some of those universes may have mutations which take them back down on to the plain, but each member of the next generation that stays on top of the hill will also have ten progeny. The result is exponential growth that produces a huge number of universes populating the ten-progeny hill, and all the other ten-progeny hills in the landscape, greatly overwhelming the number of one-progeny universes on the plain below.

In a more realistic landscape, the ten-progeny hill would just be in the foothills of a much bigger mountain, equivalent to Dawkins' Mount Improbable, with universes like our own at its top. Extending his argument to more complex universes in this way, Smolin reasons that even if you start with a single quantum fluctuation in an inhospitable part of the cosmic landscape, you will end up with a profusion of universes evolved to be good at making black holes – and, of course, there is no reason to think that there is only one quantum fluctuation in the landscape.

This leads him to suggest, on the basis that any universe chosen at random from the Multiverse must be one of the most common kind, that the laws of physics in our Universe must be close to the optimum values for the formation of black holes, and that 'most changes in the parameters of the laws of physics will decrease the rate at which black holes are produced in our universe.' This is a controversial claim, which is still the subject of fierce debate, but I can give one example in support of Smolin's argument, which also makes clear his proposal that life is a by-product of black hole formation. It involves the most famous cosmic coincidence of them all, Hoyle's anthropic insight concerning the nature of carbon nuclei.

A NEW PERSPECTIVE

In the Universe today, black holes are produced in large numbers in the death throes of stars. Not every star ends its life as a black hole – that is one of the contentious features of Smolin's argument – but astronomers estimate that there is at least one black hole, produced in previous generations of stars, for every ten thousand stars visible in the Universe today. So in a galaxy like our Milky Way, which contains a few hundred billion stars, there are tens of millions of black holes, at a minimum, with more being manufactured all the time as stars run through their life cycles. With at least a hundred billion galaxies like the Milky Way in the visible Universe, this means that as a conservative estimate there are 10^{18} to 10^{19} black holes in the visible Universe; each one, if Smolin is correct, forms the end of an umbilical cord linking our Universe to another. In order to make all those black holes and baby universes, you first need to make stars, and that is where carbon is so important.

When a cloud of gas and dust in space begins to collapse under its own weight, beginning the first stage of star formation, it gets hot, as gravitational energy is released, and the heat produces a pressure that tends to hold the cloud up and stop it collapsing. The first stars formed in the Universe, just after the Big Bang, were made from clouds containing only hydrogen and helium set in the framework of dark matter; they must have been very large, because only large, massive clouds can collapse in spite of this heating. But in this first generation of stars, and in subsequent generations, heavier elements were made, and as a result star formation became much easier. The clouds in which new stars form today contain, along with copious quantities of hydrogen and helium, traces of organic molecules such as carbon monoxide, which are good at absorbing heat from the heart of the cloud and then re-radiating it outward at infrared wavelengths which can escape into space. The clouds also contain grains of material, about as substantial as the particles in cigarette smoke, also built around carbon atoms, which shield the hearts of the clouds from the heat of any nearby starts, which might otherwise make the clouds evaporate before they could collapse.

Without going into the details, what matters is that because of its role in aiding the collapse of interstellar gas clouds, carbon is a key ingredient in the process of star formation. This means that any change in the constants of nature that makes it harder to make carbon would result in a universe which contained fewer stars and fewer black holes. The carbon resonance that Fred Hoyle drew attention to still stands out as an example of a cosmic coincidence that needs to be explained; but the explanation Smolin offers is that universes in which it is easy to make carbon will be selected from the cosmic landscape because that encourages the formation of black holes and baby universes. Our existence has nothing to do with it.

Smolin also tackles the puzzle of the other cosmic coincidences from the same perspective, showing that there are many ways to make changes in the parameters of physics that would lead to a universe in which black holes were far less common. These are the same coincidences used by proponents of anthropic cosmology to argue that the Universe we live in has been selected from the cosmic landscape by our presence, because if those coincidences did not hold we would not be here to notice them. But from Smolin's perspective our kind of life is just a by-product of the formation of stars and black holes, and it is a genuine coincidence that what is good for black holes – for example, carbon – is also good for life.*

Smolin also draws attention to a similarity between the cosmic landscape and the fitness landscape coded for by DNA, the molecule of life. DNA can code for all the forms of life on Earth, but it can also code for many other forms of life that do not, in fact, exist at present. There is a kind of DNA landscape, and 'the set of all possible

---

* 'Coincidence' is not quite the right word, because if Smolin is correct then life has evolved as a kind of parasite, taking advantage of the conditions which encourage the formation of black holes. Either way, it's black holes that really matter.

DNA sequences, and all possible collections of them, may be said to exist timelessly, as possibilities.' This resembles the way in which the set of all possible universes allowed for by string theory exists timelessly, in the cosmic landscape. 'It might then be argued,' says Smolin, 'that natural selection does not create novelty, it merely selects from a list of possibilities that always exist.'*

The similarities between natural selection at work among the forms of life on Earth and natural selection at work among the forms of universes in the cosmic landscape is even closer than you might think if you are familiar with ideas such as the 'struggle for survival' and competition for resources which winnows forms of life on Earth and leads to the 'survival of the fittest'. This struggle is extremely good at tailoring a species to fit its ecological niche, but it is variation that enables new forms of life to conquer new niches. What matters at a fundamental level is the number of offspring that an individual leaves, or more technically the differential survival rate of the organisms.† Competition for resources and the struggle for survival are important in determining the differential survival rate of forms of life on Earth, but this doesn't mean that they are the whole story, or even that they are essential to evolution. It is *variation* that is essential. As long as there are small changes from one generation to the next and the rate of reproduction in succeeding generations is related to the genetic material (for life on Earth) or to the parameters of physics (for universes), evolution will produce larger numbers of some variations on the theme than others. Smolin sums it up neatly:

While popular accounts of evolution have often stressed competition, looking at the plethora of different ways species have invented to live, it seems that an important theme of evolution might instead be the ability of the process to invent new ways of living, in order to *minimize* the actual competition among the different species.‡

* It is only fair to point out that Smolin doesn't particularly like this argument; but I do!
† Strictly speaking, this means the survival rate in many succeeding generations; it's no good having lots of offspring if they all die without reproducing. But evolutionary biologists use the term 'offspring' to cover all of that.
‡ My emphasis.

My own way of looking at this is that humankind, supposedly the pinnacle of evolution, is actually descended from a long line of losers. The fish that were good at being fish stayed in the sea, while fish that were not so good at being fish had to evolve the ability to breathe air and move on to the land; amphibians that were good at being amphibians stayed near the water, while amphibians that were bad at being amphibians, instead of competing with their relations, evolved the ability to move away from water and find a new way of life inland, and so on. Farther down the line, ape-like creatures that were good at living in trees stayed in the trees, while the unsuccessful tree-dwellers evolved and developed a different lifestyle enabling them to make a living on the plains of Africa. In terms of survival of the fittest, the most successful form of life on Earth is the one with the best survival rate, the simple bacteria that have been around essentially unchanged for more than four billion years, nearly a third of the time since the Big Bang; but exploring new parts of the DNA landscape is just as important in evolution as becoming more finely tuned to life in one particular spot.

The idea of natural selection of universes removes life, including human life, from centre stage in the story, and offers what from one point of view is a rather bleak perspective on our existence, as no more than parasites taking advantage of the way the Universe has evolved to produce a proliferation of black holes. If you feel that way, you can take comfort from the fact that Smolin's idea is by no means widely accepted, and there are counter-arguments to his reasoning, centred around the idea that it may be possible to imagine universes in which the production of black holes is even more efficient than it is in our universe. By Smolin's own reasoning that would mean that we live in a relatively rare part of the landscape, with all the problems that implies. The debate is far from over, and the issue has not been settled one way or the other. But I do not need to go into the details of the debate, because Smolin's proposal has stimulated another idea which puts life, and possibly even human life, back prominently at the centre of things.

Although natural selection has undoubtedly been at work among the varieties of life on Earth for billions of years, in the past few thousand years human beings have been doing their own selecting of individuals for characteristics we find desirable, which is why Charles Darwin had to use the term 'natural' selection to make it clear what he was talking about. By selecting the individuals we want in succeeding generations we have produced better crops and better animals for our purposes. Wheat has been developed from wild grasses; cows that yield large quantities of milk from wild cattle; the whole variety of dogs from the wolf. The whole variety of dogs is implicitly coded for in wolf DNA, exactly as in a genetic landscape, and has been teased out, or explored, by traditional methods of selection, breeding from animals with the characteristics we want to follow a path across the DNA landscape from one kind of individual animal to another.

Today, we understand so much about the genetic material that it is possible to alter DNA directly to produce modified varieties, jumping from one part of the DNA landscape to another without having to follow a path through every point in between. There is even talk of designer babies – the technology to 'improve' our children already exists, and the debate is about the desirability, or otherwise, of using this ability. So could a civilization slightly more advanced than our own do the same for universes? Instead of, or as well as, natural selection of universes, could there be artificial selection, producing designer universes, and could we be living in one of them? Could God be a gardener of universes? The answer is very much 'yes', because it is so easy to make black holes. Nature may need a star to do the trick, but technology only a tiny bit more advanced than our own could do it here on Earth. Indeed, it has been suggested that black holes might be manufactured accidentally at a collider like the LHC, an idea which underpins one of the best 'hard science' science fiction stories of recent times.*

MAKERS OF UNIVERSES

Anything can be made into a black hole if it is squeezed hard enough. For any mass, there is a critical radius, called the Schwarzschild radius, which is the radius of a black hole with that particular mass. For the

---

* *Cosm*, by Gregory Benford.

mass of the Sun, the Schwarzschild radius is about 3 kilometres, which means that if all the mass of the Sun were squeezed into a sphere 3 km across it would become a black hole, cut off from our view, with everything inside plunging towards a singularity, or into the quantum tunnel leading to a new universe. If you had as much mass as a million Suns, which corresponds to a rather small galaxy, it would only have to be contained in a sphere with a radius of 3 million km to become a black hole. But even the Earth would become a black hole if it were squeezed into a sphere with a radius of just 1 centimetre. The radius of a black hole is directly proportional to its mass. In the colliding beams of a particle accelerator like the LHC, tiny masses are squeezed together into tiny volumes. Just possibly, they may be squeezed so much that they would form tiny black holes. But, because of the negativity of gravity, it doesn't matter how small the black hole is, it still has the potential to inflate and expand away in its own dimensions to become a fully-fledged universe like our own.

One of the people who has investigated this idea in detail is Alan Guth, the father of inflation. Working with colleagues at MIT, in the late twentieth century he worked out the technicalities of what he refers to as 'the creation of universes in the laboratory'; these ideas are discussed near the end of his book *The Inflationary Universe*. His conclusion is that the laws of physics do indeed allow, in principle, that a sufficiently advanced technological civilization could create a universe, or more than one universe, in this way; the rest, Guth says, slightly tongue in cheek, is 'a mere engineering problem'. In a further refinement of the idea, it seems likely that any tiny black hole formed in this way will rapidly evaporate through the Hawking radiation process and disappear from our Universe, severing the umbilical link with the baby universe.

In the mid 1990s, Ted Harrison, a cosmologist working at the University of Massachusetts, combined elements of Smolin's idea of natural selection of universes with elements of Guth's idea of creating universes in the laboratory to come up with a scenario for the artificial selection of universes.* This is the culmination (so far) of a long

---

* Confusingly, Harrison calls this 'natural selection of universes'. But he doesn't mean the same thing that Smolin calls by the same name.

history of speculation on this topic. In 1584, Giordano Bruno outraged the established Church by suggesting that 'the excellence of God' might be 'magnified and the greatness of his kingdom made manifest [if] he is glorified not in one, but in countless suns; not in a single earth, but in a thousand, I say, in an infinity of worlds.' In 1779, the philosopher David Hume caused slightly less offence by suggesting that God might not have got it right first time, and that one universe after another 'might have been botched and bungled throughout an eternity ere this system was struck out;* much labour lost, many fruitless trials made, and a slow but continual improvement carried out during infinite ages in the art of worldmaking.' Commenting with approval on Hume's idea, Harrison asks: why stop at our Universe? Why can't the process continue, to make universes even more splendid, and perhaps even better suited to life, than the one we inhabit, whether or not you invoke God as part of the process?

Olaf Stapledon described such a process in his early work of science fiction, *Star Maker*, first published in 1937. His dreaming character watches the Star Maker at work:

In vain my fatigued, my tortured attention strained to follow the increasingly subtle creations which, according to my dream, the Star Maker conceived. Cosmos after cosmos issued from his fervent imagination, each one with a distinctive spirit infinitely diversified, each in its fullest attainment more awakened than the last; but each one less comprehensible to me . . . I strained my fainting intelligence to capture something of the form of the ultimate cosmos. With mingled admiration and protest I haltingly glimpsed the final subtleties of world and flesh and spirit, and of the community of those most diverse and individual beings awakened to full self-knowledge and mutual insight.

In one of the Star Maker's universes:

Whenever a creature was faced with several possible courses of action, it took them all, thereby creating many distinct temporal dimensions and distinct histories of the cosmos. Since in every evolutionary sequence of the cosmos there were many creatures and each was constantly faced with many possible

* 'Struck out' in the sense of coins being struck at a mint; not in the modern sense of being deleted.

courses, and the combinations of all their courses were innumerable, an infinity of distinct universes exfoliated from every moment of every temporal sequence in this cosmos.

Here is the many worlds idea, admittedly without any scientific underpinning, two decades before Hugh Everett's version!*

So – was there a Star Maker? Do we live in a designer universe?

EVOLUTION IN DESIGNER UNIVERSES

There's a problem about using the words 'intelligence' and 'design' in the same sentence in a book about the Universe. The problem is that there is a vociferous group of people, mostly based in the United States of America, who do not accept the fact of evolution, let alone the theory of natural selection put forward by Charles Darwin and Alfred Russel Wallace to explain how evolution works. They believe in the literal word of the Bible (or at least, those parts of the Bible it suits them to believe), and that each species was created, or designed, by God. They call this idea 'Intelligent Design', or ID. So in order to avoid any confusion, I need to spell out that this is *not* what I am talking about when I refer to designer universes, or to the possibility that our Universe was made deliberately by a member or members of a technologically advanced civilization in another part of the Multiverse. Such a designer may have been responsible for the Big Bang; but this still means that evolution by natural selection and all the other processes that produced our planet and the life on it have been at work in our Universe since the Big Bang, with no need for outside intervention.

Evolution is a fact, like the fact that apples fall off trees. This was already well-known in Darwin's day – indeed, his grandfather, Erasmus Darwin, was one of the earlier thinkers who puzzled over the fact of evolution before Charles Darwin was even born, and tried to find a mechanism to account for it. The mechanism that does account for it is natural selection, which was hit upon independently by Darwin and Wallace from their separate studies of the proliferation

* A similar idea appears in Jorge Luis Borges' story 'The Garden of Forking Paths'.

of life in the tropics and the 'struggle for survival'. Natural selection is the theory that explains the fact of evolution, just as the general theory of relativity is the theory that explains the fact of gravity – the reason why, among other things, apples fall off trees.

When physicists refer to 'the theory of gravity' they mean Einstein's theory, which explains the fact of gravity; when biologists refer to 'the theory of evolution,' they mean the Darwin–Wallace theory, which explains the fact of evolution.

This terminology highlights another important feature of the scientific endeavour. People who criticize the idea of evolution, such as proponents of ID, often do so partly on the grounds that it is 'just a theory'. Leaving aside the fact that the theory is natural selection, not evolution, what matters is that they have been confused by the difference between the use of the word 'theory' in everyday language and its use in a scientific context. In everyday language, someone's half-baked idea might be described as 'just a theory' – my brother reckons that the right way to add milk to tea is to pour the milk first, but that's just his theory, and I'm entitled to my own opinion. In science, a theory is a fully baked idea that has been tested by experiment and observation, and passed those tests.

Even if, or when, it fails a test, a successful theory need not be completely discarded, because any new theory that supersedes it must pass all the tests the older theory passed, as well as the new tests. In this way, Newton's theory of gravity did not become irrelevant when Einstein's theory of gravity came along. Newton's theory still works well for describing the way apples fall from trees; Einstein's theory also explains that, but in addition it explains details of, for example, the orbit of the planet Mercury, which Newton's theory cannot explain. In the same way, the Darwin–Wallace theory has been refined and improved, not least by the development of an understanding of the workings of DNA, but that does not invalidate the fundamental truth they hit upon in the nineteenth century.

Natural selection has been seen at work all around us – in one of the most beautiful and apposite examples, in the populations of finches on the Galapagos Islands, where Darwin himself marvelled at what he saw. Jonathan Weiner's book *The Beak of the Finch* goes into absorbing detail on this. Natural selection has also been seen at work in laboratory experiments, with creatures such as fruit flies who have short lifespans and can be studied over many generations. There is no need for an intelligent designer to explain how we got to be the way we are, given the laws of physics that operate in our Universe. So what scope *is* there for an intelligent designer of universes?

UNIVERSES BY DESIGN

If designers of universes make new universes by manufacturing black holes, which is the only way to do it that we are aware of at present, there are three levels at which they might operate. The first is just to manufacture black holes without any attempt to influence how the laws of physics in the new universe operate. As far as the evolution of universes is concerned, this is essentially the same as Lee Smolin's scenario of a Multiverse populated by universes created from natural black holes, but with the bonus that the designers may, over the lifetime of their own universe, manufacture many more black holes and baby universes than would have been made naturally. This is the level that humankind has nearly reached already; Gregory Benford's novel *Cosm* puts the prospect in an entertaining fictional context. Intriguingly, this might imply that intelligent life is 'selected for' in the Multiverse, since universes containing intelligences that made black holes would be more common than lifeless universes.

The second level, for a slightly more advanced civilization, would involve the ability to nudge the properties of the baby universes in a certain direction. For example, it might be possible to tweak the process of black hole formation in such a way that the force of gravity is a little stronger in the baby universe than in the parent universe, without the designers being able to say exactly how much stronger it will be.

And the third level, for a very advanced civilization, would involve the ability to set precisely the parameters of physics in the baby universe, such as the exact value of the carbon resonance, thereby designing the baby universe in detail. It is at this level that we might make an analogy with designer babies – instead of tinkering with DNA to get a perfect child, a sufficiently advanced technological civilization might tinker with the laws of physics to get a perfect universe. Crucially, though, in no case – not even at the most advanced level – would it be possible for the designers to interfere with the baby universes once they had formed. From the moment of its own big bang, each universe would be on its own.

The most startling thing about making a baby universe, as I have explained, is that it is easy – far easier than simulating a universe like our own in a computer, at least at level one. Because of this, all the arguments used by proponents of the idea of faking universes apply even more forcefully to the idea of making universes. Even if the fakers are correct (and I do not believe that they are), by their own reasoning manufactured universes should far outnumber simulated universes, so that it is much more likely (exponentially more likely, in the language of the fakers) that we are living in a manufactured universe than that we are living in a computer simulation.

Harrison's proposal suggests that there is an initial* landscape of universes, in which evolution occurs naturally by Smolin's process until at least one universe emerges with intelligence at about our level. This is the seed from which intelligent design† plus evolution leads to such a proliferation of universes like our own (in the sense of being suitable for intelligent life) that 'unintelligent' universes become a tiny fraction of the whole Multiverse. The first intelligent universe may be produced by chance, but thereafter manufactured universes proliferate and dominate the scene. In that sense, it seems likely that the existence of the Universe we see around us may be a result of *both* anthropic selection (extending the term 'anthropic' to refer to any intelligent life forms) *and* tailoring to suit life, within the context of the cosmic landscape of the Multiverse.

As Harrison put it in an article published in the *Quarterly Journal of the Royal Astronomical Society*:

Life itself takes over the creation business ... the superior beings who created our universe inhabited a universe not greatly unlike our own. They

---

* I'm doubtful about using the word 'initial', since as I discussed earlier the flow of time may be an illusion; but everyday language isn't really up to the task of describing a timeless Multiverse, so it will have to do.

† *Not* the kind of 'Intelligent Design' proposed by anti-evolutionists!

were not only intelligent but intelligible, and were perhaps similar to our distant descendants who might also create universes. How these superior beings created our universe and how their own was created now become comprehensible issues open to inquiry.

The intelligence required to do the job may be superior to ours, but it is a finite intelligence reasonably similar to our own, not an infinite and incomprehensible God. The most likely reason for such an intelligence to make universes is the same as the reason why people do things like climbing mountains or studying the nature of subatomic particles using accelerators like the LHC – because they can. A civilization that has the technology to make baby universes might well find the temptation irresistible, while at the higher levels of universe design, if the superior intelligences are anything at all like us there would be an overwhelming temptation to improve upon the design of their own universes.

This provides the best resolution yet to the puzzle Albert Einstein used to raise, that 'the most incomprehensible thing about the Universe is that it is comprehensible.' The Universe is comprehensible to the human mind because it was designed, at least to some extent, by intelligent beings with minds similar to our own. Fred Hoyle put it slightly differently. 'The Universe,' he used to say, 'is a put-up job.' I believe that he was right. But in order for that 'put-up job' to be understood, we need all of the elements discussed in this book.

I have argued that a comparison of 'causal patch' arguments and thermodynamics implies that although nothing outside our causal patch can ever affect us, we are only here because everything outside our causal patch exists. The very fact that we exist seems to be the best evidence available that we do indeed live in a Multiverse. The best mathematical description of that Multiverse that we have today is the string landscape, which Leonard Susskind has shown to be essentially the same as the Many Worlds 'landscape' of Hugh Everett, an idea expressed most clearly in recent times by David Deutsch. From either perspective, Ted Harrison's refinement of Lee Smolin's idea of the evolution of universes to include the role of intelligent designers of universes completes the picture. There is no puzzle about the cosmic coincidences after all. The Universe was indeed set up to provide a home for life; but once the Universe got started, life evolved through a process of natural selection with no need for outside interference. It isn't so much that Man was created in God's image, but that the Universe was created in the image, more or less, of the universe of the Designers.

Further Reading

MORE OF MY OWN THOUGHTS

Although I did not set out as a writer with the intention of providing an overview of our place in the Universe (or universes), with hindsight most of my* books fit that overall theme. So it may be of interest to set some of them in that context, thematically rather than in the order of publication.

Science: A History (Allen Lane, 2002) provides an overview of how humankind developed an understanding of the world around us. *In Search of the Big Bang* (Penguin, 1998) deals with the origin of our Universe, while *The Universe: A Biography* (Allen Lane, 2007) brings the story up to date and *Stardust* (Allen Lane, 2000) puts life on Earth in our cosmic context. *Deep Simplicity* (Allen Lane, 2004) discusses the emergence of complexity in a Universe governed by simple rules, and the emergence of our kind of complexity in the context of evolution by natural selection is covered in *Being Human* (Dent, 1993). Evolution in a broader sense is a theme of our biography of James Lovelock, *He Knew He Was Right* (Allen Lane, 2009). The need to set all of this in a framework allowing for the existence of other universes is developed from the viewpoint of quantum theory in *In Search of Schrödinger's Cat* (Corgi, 1985) and *Schrödinger's Kittens* (Weidenfeld & Nicolson, 1995), and in a more explicitly 'anthropic' context in a collaboration with Martin Rees, *The Stuff of the Universe* (Penguin, 1995). The need for more dimensions than the familiar three of space and one of time, part of the quest for a 'theory of everything', is covered in *In Search of Superstrings* (Icon, 2007). And *Q is for Quantum* (Weidenfeld & Nicolson, 1998) provides an A-to-Z guide to the quantum world.

* I should say 'our' books, since many of them involved collaborations with Mary Gribbin.

OTHER BIBLIOGRAPHY

Books marked with an asterisk are more technical but particularly relevant; books marked + are fiction, but use the Multiverse idea in one way or another.

Amir Aczel, *Entanglement*, Four Walls Eight Windows, New York, 2002.

+Poul Anderson, *Time Patrol*, Baen, New York, new edition, 2006.

Svante Arrhenius, *Worlds in the Making*, Harper, London, 1908.

Jim Baggott, *Beyond Measure*, Oxford UP, 2004.

Jim Baggott, *A Beginner's Guide to Reality*, Penguin, London, 2005.

Julian Barbour, *The End of Time*, Weidenfeld & Nicolson, London, 1999.

John Barrow, *The Infinite Book*, Cape, London, 2005.

John Barrow, *New Theories of Everything*, Oxford UP, 2007.

*John Barrow and Frank Tipler, *The Anthropic Cosmological Principle*, Clarendon Press, Oxford, 1986.

*John Barrow, Paul Davies and Charles Harper (editors), *Science and Ultimate Reality*, Cambridge UP, 2004.

+Gregory Benford, *Cosm*, Orbit, London, 1998.

*Niels Bohr, *Atomic Theory and the Description of Nature*, Cambridge UP, 1934.

*Ludwig Boltzmann, *Lectures on Gas Theory*, University of California Press, Berkeley, 1964; originally published in two volumes in German, in 1896 and 1898.

+Jorge Luis Borges, *Collected Fictions*, trans. Andrew Hurley, Allen Lane, 1998.

Nick Bostrom, *Anthropic Bias*, Routledge, London, 2003.

*Julian Brown, *Minds, Machines, and the Multiverse*, Simon & Schuster, New York, 2000.

Giordano Bruno, *Gesammelte Werke*, edited by L. Kuhlenbeck, Eugen Diederichs, Jena, in five volumes, 1904–7.

*Bernard Carr (editor), *Universe or Multiverse?*, Cambridge UP, 2007.

+Lewis Carroll, *The Annotated Alice*, edited by Martin Gardner, Penguin, London, 2001.

+Arthur C. Clarke, *The Other Side of the Sky*, Harcourt, New York, 1958.

*J. Cornell (editor), *Bubbles, Voids, and Bumps in Time: The New Cosmology*, Cambridge UP, 1989.

Peter Coveney and Roger Highfield, *The Arrow of Time*, W. H. Allen, London, 1990.

Charles Darwin, *The Origin of Species*, Penguin, London, 1968 (reprint of the first edition, published by John Murray in 1859).

*Paul Davies, *The Accidental Universe*, Cambridge UP, 1982.

Paul Davies, *About Time*, Viking, London, 1995.

Paul Davies and Julian Brown (editors), *The Ghost in the Atom*, Cambridge UP, 1986.

Richard Dawkins, *Climbing Mount Improbable*, Viking, London, 1996.

David Deutsch, *The Fabric of Reality*, Allen Lane, London, 1997.

*Bryce DeWitt and Neil Graham (editors), *The Many-Worlds Interpretation of Quantum Mechanics*, Princeton UP, 1973.

+Philip K. Dick, *Counter-Clock World*, Berkley Medallion, New York, 1967.

+Philip K. Dick. *The Man in the High Castle*, Gollancz, London, 1975.

Arthur Eddington, *The Nature of the Physical World*, Cambridge UP, 1928.

Lewis Carroll Epstein, *Relativity Visualized*, Insight Press, San Francisco, revised edition, 1987.

Richard Feynman, *The Character of Physical Law*, MIT Press, Cambridge, Mass., 1967.

Richard Feynman, *Six Easy Pieces*, Addison-Wesley, Massachusetts, 1995.

Galileo Galilei, *Two New Sciences*, trans. Stillman Drake, University of Wisconsin Press, Madison, 1974; originally published in Italian in 1638.

George Gamow, *My World Line*, Viking, New York, 1970.

+David Gerrold, *The Man Who Folded Himself*, BenBella, New York, new edition, 2003.

Donald Goldsmith, *The Runaway Universe*, Perseus Books, Cambridge, Mass., 2000.

Brian Greene, *The Elegant Universe*, Norton, New York, 1999.

+John Gribbin, *TimeSwitch*, PS Publishing, London, 2009.

John Gribbin and Mary Gribbin, *Time Travel for Beginners*, Hodder/Headway, London 2008.

Alan Guth, *The Inflationary Universe*, Cape, London, 1997.

Edward Harrison, *Masks of the Universe*, second edition, Cambridge UP, 2003.

Lawrence Henderson, *The Fitness of the Environment*, Macmillan, New York, 1913. (Reprinted in 1970 by Peter Smith, Gloucester, Mass.)

Nick Herbert, *Quantum Reality*, Rider, London, 1985.

Tony Hey and Patrick Walters, *The Quantum Universe*, Cambridge UP, 1987.

Banesh Hoffman, *Albert Einstein*, Viking, New York, 1972.

Fred Hoyle, *Galaxies, Nuclei, and Quasars*, Heinemann, London, 1965.

+Fred Hoyle, *October the First is Too Late*, Heinemann, London, 1966.

David Hume, *Dialogue Concerning Natural Religion*, edited by J. V. Price, Oxford UP, 1976; original published in 1779.

James Jeans, *The Mysterious Universe*, Cambridge UP, 1930.

George Johnson, *A Shortcut Through Time*, Cape, London, 2003.

Michio Kaku, *Parallel Worlds*, Allen Lane, London, 2005.

Janna Levin, *How the Universe Got Its Spots*, Weidenfeld & Nicolson, London, 2002.

Seth Lloyd, *Programming the Universe*, Random House, New York, 2005.

*Malcolm Longair (editor), *The Large, the Small, and the Human Mind*, Cambridge UP, 1997.

James Lovelock, *The Ages of Gaia*, Oxford UP, 1988.

*W. H. McCrea and M. J. Rees, *The Constants of Nature*, Royal Society, London, 1983.

Walter Moore, *Schrödinger: Life and Thought*, Cambridge UP, 1989.

+Ward Moore, *Bring the Jubilee*, Equinox/Avon, New York, 1976.

Heinz Pagels, *The Cosmic Code*, Simon & Schuster, New York, 1982.

William Paley, *Natural Theology*, edited by Matthew Eddy and David Knight, Oxford UP, 2006. (First published in 1802; in book form in *The Works of William Paley*, ed. R. Lynam, 1825.)

C. F. A. Pantin, in *Biology and Personality*, edited by I. T. Ramsay, Blackwell, Oxford, 1965.

C. F. A. Pantin, *The Relations Between the Sciences*, Cambridge UP, 1968.

+Fred Pohl, *The Coming of the Quantum Cats*, Bantam, New York, 1986.

*Huw Price, *Time's Arrow and Archimedes' Point*, Oxford UP, 1996.

Ilya Prigogine and Isabelle Stengers, *Order Out of Chaos*, Heinemann, London, 1984.

Lisa Randall, *Warped Passages*, Penguin, London, 2006.

Martin Rees, *Before the Beginning*, Simon & Schuster, London, 1997.

Martin Rees, *Just Six Numbers*, Weidenfeld & Nicolson, London, 1999.

Eugen Shikhovtsev, *Biographical Sketch of Hugh Everett III*, trans. Kenneth Ford. At: http://space.mit.edu/home/tegmark/everett/everett.html

Andrei Sakharov, *Alarm and Hope*, Knopf, New York, 1978.

D. W. Singer, *Giordano Bruno*, Schuman, New York, 1950.

Lee Smolin, *The Life of the Cosmos*, Oxford UP, New York, 1997.

*Lee Smolin, *Three Roads to Quantum Gravity*, Weidenfeld & Nicolson, London, 2000.

+Olaf Stapledon, *Last and First Men, and Star Maker* (combined reprint), Dover, New York, 1968.

Victor Stenger, *Timeless Reality*, Prometheus, New York, 2000.

Victor Stenger, *The Comprehensible Cosmos*, Prometheus, New York, 2006.

Paul Steinhardt and Neil Turok, *Endless Universe*, Doubleday, New York, 2007.

Leonard Susskind, *The Cosmic Landscape*, Back Bay Books, New York, 2006.

*Kip Thorne, *Black Holes and Time Warps*, Picador, London, 1994.

*Richard Tolman, *Relativity, Thermodynamics, and Cosmology*, Clarendon Press, Oxford, 1934.

Alex Vilenkin, *Many Worlds in One*, Hill & Wang, New York, 2006.

Alfred Russel Wallace, *Man's Place in the Universe*, McClure, Phillips & Co., New York, 1903.

Jonathan Weiner, *The Beak of the Finch*, Cape, London, 1994.

+H. G. Wells, *The Time Machine*, Penguin Classics, London, 2005 (originally published in 1895).

*John Wheeler, 'Beyond the end of time', in *Black Holes, Gravitational Waves and Cosmology*, edited by Martin Rees, Remo Ruffini and John Wheeler, Gordon & Breach, New York, 1974.

John Wheeler and Kenneth Ford, *Geons, Black Holes and Quantum Foam*, Norton, New York, revised edition, 2000.

*John Wheeler and Wojciech Zurek (editors), *Quantum Theory and Measurement*, Princeton UP, 1983.

Gerald Whitrow, *The Structure and Evolution of the Universe*, Harper & Row, New York, 1959.

Glossary

anthropic cosmology   The argument that the Universe we see has to be the way it is, or we would not be here to see it. This does not necessarily mean that the Universe has been 'designed' with us in mind; if there are many universes, then life forms like us will only exist in those universes where life forms like us can exist. In the same way, the oceans are not 'designed' for fish; fish have evolved to fit their environment. But any fishy philosophers would realise that fish can only live in water, so it is unsurprising to find that there is water all around them.

antimatter   A kind of mirror image version of the kind of *matter* we are made of, with reversed properties. For example, an anti-*electron* (also known as a *positron*) has positive electric charge instead of negative electric charge.

antiparticle   A particle of *antimatter*.

astronomical unit (AU)   Distance measurement used in astronomy. 1 AU is equivalent to the distance from the Earth to the *Sun* (149,597,870 km).

atom   The smallest component of everyday *matter* that takes part in chemical reactions. All *elements*, such as oxygen or iron, are made of particular kinds of atoms. Each atom is made up of a tiny central *nucleus* surrounded by a cloud of *electrons*. There is one electron in the cloud for every proton in the nucleus.

background radiation   See *cosmic background radiation*.

baryon   A particle of *baryonic matter*.

baryonic matter   Name given to *matter* like the everyday matter here on Earth, made of *protons*, *neutrons* and *electrons*. Strictly speaking, electrons are not baryons, but their *mass* is tiny compared with that of protons and neutrons.

Big Bang   Popular term for the origin of our visible *Universe* in a hot, dense fireball some 13.7 billion years ago. By extension, other *universes* may have started in their own big bangs.

Big Bang nucleosynthesis   See *nucleosynthesis*.

big bounce   The idea that instead of ending in a *big crunch* a collapsing *universe* might 'bounce' into a new *big bang*.

big crunch   Popular term for the hypothetical death of a *universe* in a final collapse, like the *Big Bang* in reverse. It is now clear that our *Universe* will not end in this way.

billion   One thousand million.

binary system   A pair of *stars*, or a star and a *black hole*, orbiting around one another.

bit   The smallest piece of information, equivalent to a switch being either on or off. Short for binary digit.

black body   An object that both perfectly absorbs *electromagnetic radiation and* perfectly emits electromagnetic radiation.

black body radiation   The radiation emitted by a *black body*.

black hole   A region of *spacetime* bent round on itself by gravity so that nothing, not even light, can escape. See *Event horizon*.

block universe   Model of reality which says that all moments in time exist in *spacetime* just as all positions in space exist – that the year 1452, or 3173, is just as real as today even though we are not experiencing it, in the same way that New York and Mumbai are just as real as London, even if you are living in London.

blueshift   See *Doppler effect*.

Boltzmann fluctuation   The temporary appearance of a patch of order in a system in *thermodynamic* equilibrium as a result of the chance motion of *atoms* and *molecules*.

brane   A term derived from the word 'membrane' to refer to surfaces in any number of dimensions. A string is a 1-brane, a two-dimensional sheet is a 2-brane, and so on.

byte   Eight *bits*.

causal patch   Any region of space, surrounding an object, small enough for light signals to have crossed the region since the *Big Bang*. The object at the centre of such a region has had time to influence other objects in the patch since the Big Bang, but it cannot yet have caused anything to happen outside that patch. Equally, nothing outside the patch can have influenced that object since the Big Bang.

cepheid   A kind of variable *star* that changes brightness in a regular way that enables astronomers to work out its average brightness and therefore how far away it is.

CERN   The European particle accelerator laboratory near Geneva, where beams of particles are collided head-on in 'atom smashing' experiments. Among many other achievements, these experiments have proved that *time dilation* and the *twin effect* are real.

Chandrasekhar limit   The maximum *mass* for a *white dwarf*, equivalent to 1.4 times the *mass* of our *Sun*.

classical physics   The rules and equations that apply to things much bigger than *atoms*.

cold dark matter (CDM)   Another name for *dark matter*.

collapse of the wave function   The idea, central to the *Copenhagen Interpretation*, that *quantum* entities such as *electrons* exist as waves until they are observed, then 'collapse' into point-like particles.

compactification   The rolling up of the 'extra' dimensions of space required by *string theory* to make them invisible.

complementarity   The feature of *quantum physics* which says that a single model may be inadequate as a description of reality. For example, an *electron* can be described as either a wave or a particle, depending on the circumstances. These are said to be complementary properties of the electron.

conjugate variables   Pairs of properties, such as position/momentum and energy/time, linked by quantum *uncertainty*.

Copenhagen Interpretation   One way of using the equations of *quantum mechanics* to predict the outcome of experiments. This was the standard 'explanation' of the quantum world from the 1930s to the 1980s, but raises as many questions (such as the *collapse of the wave function*) as it answers.

cosmic background radiation   Radiation left over from the *Big Bang*, detectable today in the form of a weak hiss of radio noise coming from all directions in space. This is almost perfect *black body radiation.*

Cosmic Landscape   see *string landscape.*

cosmic rays   Energetic particles from space that hit the Earth's atmosphere.

cosmological constant   A measure of the amount of *dark energy.*

cosmological redshift   Stretching of light from a distant object to longer wavelengths caused by the expansion of the *Universe.* Not the same as the *Doppler Effect.*

critical density   The density for which the *spacetime* of the *Universe* is flat. The critical density is equivalent to the presence of about five hydrogen *atoms* in every cubic metre of space.

dark energy   A form of energy which fills all of space, detected only by its influence on the way the *Universe* expands.

dark matter   Material detected only by its gravitational pull, which affects the way galaxies move and how the Universe expands. There is six times more dark matter than there is *baryonic matter.*

D-brane   A surface on which the ends of the open strings of *string theory* can attach. See *brane.*

decoherence   The way *quantum* interference effects get spread out among larger groups of particles in the outside environment, so that quantum information is spread out more widely and lost in the noise of everything else that is going on.

de Sitter space   Mathematical description of an eternally expanding *universe.* See *Eternal inflation.*

Doppler effect   A change in the wavelength of light or sound which happens when the object emitting the waves is moving towards or away from the observer. Light from an object moving towards you is squashed to shorter wavelengths (blueshift); light from an object moving away from you is stretched to longer wavelengths (redshift). See also *Cosmological redshift.*

Einstein–Rosen bridge   Another name for a *wormhole.*

electromagnetic radiation   Any form of radiation, including light, radio waves and X-rays, that is made up of electricity and magnetism, described by *Maxwell's equations.*

electromagnetism　Description of electricity and magnetism in one mathematical package, described by *Maxwell's equations*.

electroweak theory　Description of *electromagnetism* and the *weak force* in one mathematical package. A step towards a *grand unified theory*.

electron　The lightest of the building blocks of *matter*, an electrically negative particle found in the outer part of an *atom*. Electrons can also behave as waves.

electron volt (eV)　Unit of energy used by physicists. 1 eV is the energy gained by a single *electron* when it is accelerated across an electric potential difference of 1 volt. A 100-watt light bulb burns energy at the rate of 624 billion billion eV per second.

element　See *atom*.

entropy　A measure of the amount of (dis)order in a system. The higher the entropy, the less organised and more random the system is; the lower the entropy, the more organised it is, with a higher information content.

escape velocity　The minimum speed necessary for an object to escape from the gravitational grip of another object. The escape velocity from the surface of the Earth is 11.2 km per second. The escape velocity from within the *event horizon* of a *black hole* exceeds the *speed of light*, which is why nothing can escape.

eternal inflation　Exponentially expanding space within which *quantum fluctuations* form new *universes* that populate the *Cosmic Landscape*.

event horizon　The imaginary surface surrounding a *black hole* which marks the limit from within which nothing can escape. The horizon is imaginary in the sense that there is no physical marker in space to reveal its presence, just as there is no natural marker on the surface of the Earth to tell you where the equator is. But it is real in the sense that once anything crosses the event horizon from outside it can never escape – as if, once you stepped across the equator from the Northern Hemisphere to the Southern Hemisphere you could never return to the North – and you would never know until you tried.

exponential notation　A way of writing numbers in terms of powers of (usually) 10. 10^2 is 100, 10^3 is 1,000, and so on, where the numbers 2, 3, and so on are the exponents. This is particularly useful for writing large numbers such as 10^{36}, which means a 1 followed by 36 zeroes. You can also represent small numbers using negative exponents. 10^{-1} is 0.1, 10^{-2} is 0.01, and so on.

In these examples, 10 is the base. Any number can in principle be used as the base; in computing, base 2 is common, and a kilobyte is 2^{10} bytes (which is actually 1,024 bytes, not exactly 1,000 bytes).

extension   The equivalent of length in four dimensions, involving time as well as the three dimensions of space.

false vacuum   A state of 'empty space' which has more energy than another state of empty space. It is a 'false' vacuum because it is possible for this vacuum to decay into the lower state, giving up energy as it does so. Such a process may power *inflation*.

Feynman diagram   A special kind of *spacetime diagram* which portrays the way particles like *electrons* and *protons* interact with each other by exchanging *photons*.

field   The region in which an object exerts a force on another object without touching it. The most familiar example is a magnetic field. When a simple bar magnet is placed under a sheet of paper and iron filings are sprinkled on the paper, the filings line up along the 'lines of force' which reveal the presence of the field.

galaxy (small 'g')   A large island in space containing many *stars* – up to hundreds of billions of stars like the *Sun*.

Galaxy (capital 'G')   Our home *galaxy*, also known as the Milky Way.

gamma ray   A very energetic *photon*.

general theory of relativity   The theory, developed by Albert Einstein, that describes the relationship between *matter* and *gravity* in terms of curved *spacetime*.

Grand Unified Theory (GUT)   Any theory that attempts to describe all the forces of nature except *gravity* in one mathematical package.

graviton   The particle that carries the gravitational force.

gravity   A force exerted by any object with *mass* on any other object with mass. The size of the force depends on the mass of the object divided by the square of the distance from it. This is called the *inverse square law* of gravity.

heat death of the Universe   The idea that one day all the *stars* will burn out and the temperature everywhere in the *Universe* will be the same.

horizon   See *Event horizon*.

Hubble constant   Also known as the Hubble parameter, a number which measures how fast the *Universe* is expanding.

Hubble parameter   See *Hubble constant*.

inertia   See *mass*.

inflation   The early phase in the development of the *Universe* when a tiny *quantum fluctuation* expanded to about the size of a grapefruit in a tiny fraction of a second.

infrared   A form of invisible light with wavelengths longer than red light. See *Spectrum*.

interference pattern   The pattern made when two sets of waves overlap. This could be, for example, ripples on a pond, or light spreading out from two holes in a piece of cardboard.

inverse square law   Any law describing an effect, like *gravity*, which falls off as 1 divided by the square of distance. At twice the distance, the effect is one-quarter as strong, and so on.

Kelvin temperature scale   Temperature measured from the absolute zero of temperature, $-273.15\,°C$. Each degree on the Kelvin scale is the same size as a degree on the Celsius scale, but is written without the 'degrees' sign. So $0\,°C$ is the same as 273.15 K, and so on.

kinetic energy   The energy something has because it is moving.

length contraction   The way the length of a moving object shrinks along the direction of its motion.

lambda (Λ) field   Another name for *dark energy*.

landscape   see *string landscape*.

light year   The distance travelled by light in one year, 9.46 million million km. A light year is a measure of distance, not of time.

look back time   The time it has taken for light from a distant object to reach us. Because the *speed of light* is finite, we see things as they were long ago when the light left them. In this sense, telescopes are like time machines, giving us a view of the past history of the *Universe*.

main sequence star   A *star* in the quiet prime of its life, like the *Sun*.

Many Worlds Hypothesis   The idea that there are many different versions of reality, somehow lying 'next door' to each other like the pages in a book. These are sometimes referred to as *parallel universes*, or *parallel worlds*. Also known as the Many Worlds Interpretation.

Many Worlds Interpretation   See *Many Worlds Hypothesis*.

mass   A measure of the amount of *matter* something contains. It is mass that causes an object's resistance to being pushed around (*inertia*), and it is mass that produces the gravitational force which attracts objects to one another. The bigger the mass, the bigger the inertia and the bigger the gravitational pull.

matter   See *mass*.

Maxwell's equations   A set of equations that describe the way electricity and magnetism behave. In particular, the equations tell us what the *speed of light* is, because light is a form of *electromagnetic radiation*.

metaverse   Term used by some people for the *Multiverse*.

Milky Way   The name of our home *galaxy*; also called the *Galaxy*, with a capital 'G.'

Minkowski diagram   A way of representing the movement of things through *spacetime* on a kind of graph, or map.

molecule   A group of *atoms* held together in a stable unit by electromagnetic forces.

momentum   The *mass* of an object multiplied by its velocity.

M-theory   An over-arching theory that includes all the known versions of *string theory*.

Multiverse   Everything there is.

nebula   In its modern usage, a cloud of gas and dust between the *stars*. Before it was realized that the objects now known as *galaxies* lie outside the *Milky Way*, some of them, like the Andromeda galaxy, were also labelled nebulae; but as applied to galaxies the term is now obsolete.

neutron   A neutral particle found in the *nucleus* of an *atom*.

neutron star   A kind of dead *star*. Each cubic centimetre of a neutron star would have a *mass* of a hundred million tonnes. The maximum possible mass for a neutron star is about three times the mass of our *Sun*.

Newtonian physics Another name for *classical physics*.

nucleon General name for *protons* and *neutrons*.

nucleosynthesis The natural processes which build up heavier elements from lighter ones. A little of this happened in the *Big Bang* (Big Bang nucleosynthesis) but most of the elements except hydrogen and helium have been manufactured inside *stars* (stellar nucleosynthesis).

nucleus The central part of an *atom*, made up of *protons* and *neutrons*.

parallel universe See *Many Worlds Hypothesis*.

parallel worlds See *Many Worlds Hypothesis*.

parsec (pc) Distance measurement used in astronomy. 1 pc is equivalent to 3.2616 *light years*.

photon A particle of light, or of any electromagnetic field or wave.

Planck length The smallest possible size for anything, the *quantum* of size. In centimetres, roughly a decimal point followed by 32 zeroes and a 1.

Planck scale General term for anything roughly the size of the *Planck length*.

Planck time The smallest possible unit of time – the *quantum* of time. In seconds, roughly a decimal point followed by 42 zeroes and a 1.

Planck's constant A very small number, usually written as h, that relates the wavelength of a *quantum* object, such as an *electron*, to its *momentum*.

Poincaré cycle time See *Poincaré recurrence time*.

Poincaré recurrence time The time it takes, statistically speaking, for the *atoms* of a system, such as atoms of gas trapped in a box, to return to any chosen state without outside interference.

positron *Antimatter* counterpart of the *electron*.

potential energy Energy something possesses because of its position or state. If a lower energy state exists (for example, if water is in a mountain lake) the potential energy can be released by moving to the lower energy state (for example, when water flows downhill).

principle of terrestrial mediocrity The idea that we do not occupy a special place in the *Universe*. By extension, the idea that our Universe is a typical member of the *Multiverse*.

proton A positively charged particle found in the *nucleus* of an *atom*.

pulsar A spinning *neutron star* that emits pulses of radio noise, like a radio lighthouse in space.

quantum (adjective) Referring to things on the smallest scales, as in *quantum physics*.

quantum (noun) The smallest amount of anything that can exist. A *photon*, for example, is a quantum of light.

quantum computer A computer working on quantum principles. See *qubit*. The fact that quantum computers work is powerful evidence in support of the idea of the *Multiverse*.

quantum dot The 'switch' used in a *quantum computer* to store *qubits*.

quantum electrodynamics (QED) The theory of light and *matter* that explains how *atoms* and *molecules* work.

quantum field theory Mathematical description of the interactions between fundamental particles.

quantum fluctuation A (usually) short-lived bubble of energy on the *Planck scale*. But sometimes, such a bubble can grow to become a *universe*. See *inflation*. Synonymous for the present purposes with *vacuum fluctuation*.

quantum foam The frothy nature of *spacetime* on the *Planck scale*, caused by *quantum fluctuations*.

quantum gravity The long-sought theory that will explain everything in one mathematical package.

quantum mechanics Another term for *quantum physics*.

quantum physics The laws and equations that describe the way small things like *electron*s and *atoms* behave.

quantum scale See *Planck scale*.

quark Fundamental entity from which 'particles' such as *protons* and *neutrons* are made. But see *string*.

quasar The active core of a *galaxy*, fuelled by *matter* falling in to a *black hole*.

qubit Pronounced 'cubit', the equivalent of a *bit* in a *quantum computer*.

In a quantum computer, a switch can be both on and off at the same time, in a *superposition*.

recurrence time   See *Poincaré recurrence time*.

redshift   See *Cosmological redshift; Doppler Effect*.

Schrödinger's cat   A mythical cat that, according to one version of *quantum physics*, can be dead and alive at the same time.

Schwarzschild radius   The radius of the *event horizon* of a *black hole*.

second law of thermodynamics   Heat always flows from a hotter object to a cooler object, never the other way around. Scientifically speaking, this means that the *entropy* of a closed system always increases. Or in everyday language, things wear out.

singularity   A point with zero volume, or a line with zero width.

spacetime   The unification of the three dimensions of space and the fourth dimension of time, described by the *general theory of relativity*.

spectroscopy   The technique of analysing the light from an object to reveal its composition. Each *element*, such as hydrogen or carbon, produces distinctive lines in the *spectrum*, equivalent to a fingerprint or a barcode.

spacetime   description of the three dimensions of space and the dimension of time in one package. See *general theory of relativity*.

spacetime diagram   See *Minkowski diagram, Feynman diagram*.

spectroscopy   Technique for determining the composition of an object, for example, a *star* or *galaxy*, by analysing the light in its *spectrum*.

spectrum   The rainbow pattern of coloured light seen when white light is split up using a prism. The range of colours we can see extends from red through orange, yellow, green, blue and indigo to violet. Red has the longest wavelength, violet the shortest.

speed of light   The ultimate speed limit for anything moving through space, 299,792,458 metres per second (very nearly 3×10^8 m/s).

star   A hot ball of gas, many times bigger than a *planet*, which shines because energy is released by nuclear reactions going on in its interior. The *Sun* is a star.

statistical mechanics   The development of *thermodynamics* into a way of applying statistics to calculate the behaviour of systems made up of large numbers of components, such as the behaviour of a gas made of many *molecules*. In this book, used synonymously with *thermodynamics*.

Steady State Hypothesis   The idea that the *Universe* is eternal and unchanging in its overall appearance. Although this is not the case, elements of the steady state idea appear (often unacknowledged) in the modern cosmological idea of *inflation*.

stellar nucleosynthesis   See *nucleosynthesis*.

string   See *string theory*.

string landscape   Description of the *Multiverse* in terms of *string theory*.

string theory   Any one of several descriptions of the world in terms of tiny loops of vibrating *string*, rather than as point-like particles. In some versions there may also be 'open' strings.

strong nuclear force   The force that holds *protons* and *neutrons* together in *nuclei*.

Sun   The *star* at the centre of our Solar System.

supernova   The explosion of a *star* with more than a certain amount of *mass* at the end of its life. For a brief time, a supernova may shine as brightly as an entire *galaxy* of ordinary stars like the *Sun*. A supernova leaves behind a remnant in the form of either a *neutron star* or a *black hole*.

supernova 1a (Sn1a)   A particular kind of *supernova* which all have the same brightness, making them useful as distance indicators across the *Universe*. Studies of Sn1a show that the expansion of the Universe is accelerating, a result of the effect of *dark energy*.

superposition   The ability of a *quantum* system to exist in two (or more) states at the same time. See *Schrödinger's cat*.

superstring theory   Another term for *string theory*.

Theory of Everything (TOE)   Any theory that attempts to describe all the forces of nature including *gravity* in one mathematical package. See *Grand Unified Theory*, *string theory*.

thermodynamics   The scientific study of energy, *entropy* and heat. In this book, used synonymously with *statistical mechanics*.

time dilation   The way time stretches out for a moving object, so that moving clocks run slow.

twin effect If one of two identical twins goes on a long journey at nearly the *speed of light*, when he or she gets home the travelling twin will be younger than the twin who stayed at home, because of *time dilation*.

ultraviolet A form of invisible light with wavelengths shorter than violet light. See *spectrum*.

uncertainty In *quantum physics*, a precise measure of the accuracy with which each member of a pair of *conjugate variables* can be known.

universe (small 'u') The term used for mathematical or computer models of how spacetime evolves, and also for hypothetical regions of spacetime, perhaps in other dimensions, inaccessible in principle from any observations we can make. Such inaccessible regions of spacetime are part of the *Multiverse*, other components of the Multiverse equivalent to our own *Universe*.

Universe (capital 'U') The totality of everything in space and time of which we could ever, in principle, have direct knowledge – our component of the *Multiverse*.

ultraviolet A form of invisible light with wavelengths shorter than violet light. See *spectrum*.

vacuum fluctuation See *quantum fluctuation*.

velocity A measure of both the speed an object is moving at and the direction it is moving in. For example, 100 km/hour is a speed; 100 km/hour in a northwesterly direction is a velocity.

wave function Mathematical description of *quantum* entities such as *electrons* in terms of waves. Works in some circumstances but not in others. See *Copenhagen Interpretation*.

weak force Short-range force involved in radioactive decay of subatomic particles.

white dwarf A kind of dead *star*. The *Sun* will end its life as a white dwarf. One cubic centimetre of white dwarf *matter* would have a *mass* of about 1 tonne

wormhole A tunnel through *spacetime* joining two *black holes*.

X-ray An energetic *photon*.

Index

Glossary definitions are shown by 'g' after the page number